S10
S18
S21
S23

Carpentry
& Joinery

Activities

05

Carpentry & Joinery

Book 2
Practical Activities

Peter Brett

First edition of *Carpentry and Joinery for Building Craft Students* Bks 1 and
2 published in 1981 by:
Hutchinson Education
First edition of *Carpentry and Joinery for Advanced Craft Students: Site Practice*
published in 1985 by:
Hutchinson Education

Second edition (of *Carpentry and Joinery* Bk 2 incorporating material from
Site Practice) published in 2004 by:
Nelson Thornes Ltd
Delta Place
27 Bath Road
CHELTENHAM
GL53 7TH
United Kingdom

04 05 06 07 08 / 10 9 8 7 6 5 4 3 2 1

A catalogue record for this book is available from the British Library

ISBN 0 7487 8502 7

Illustrations by Peters and Zabransky; and Richard Morris
Page make-up by Florence Production Ltd, Stoodleigh, Devon

Printed and bound in Great Britain by Scotprint

Contents

Introduction   *vii*

1. Timber ground and upper floor construction   **1**

    Types of floor   1
    Timber floor construction   6
    Floor covering and finishing   25
    Insulation, preservation and finishing   29

2. Flat and pitched roof construction   **35**

    Roof types   35
    Flat roof construction   38
    Verge and eaves finishes   43
    Pitched roof construction   46
    Traditional roof structures   47
    Roof erection   60
    Prefabricated trussed rafters   62
    Verge and eaves finishing   73
    Preservation, insulation and ventilation   79
    Irregular plan roofs   81

3. Partition walls   **87**

    Timber stud partitions   87
    Metal stud partitions   99
    Proprietary panel partitions   101

4. Timber frame construction and timber engineered components   **105**

    Timber-frame construction   105
    Temporary 'knock-down' buildings   106
    Permanent timber-framed buildings   108
    Barriers   115
    Cladding   118
    Glue-laminated timber (glulam)   121
    Shell roof construction   123
    Curtain walling   125

5. Stairs   **128**

    Stair types   128
    Stairway construction and assembly   132
    Installation of stairs   143
    Protection of completed work   152

6. Doors, frames and linings   **154**

    Introduction   154
    Door performance requirements   156
    Door types   157
    Matchboarded doors   158

Panelled and glazed doors   164
Flush doors   171
Other types of door   175
Methods of door operation   181
Door ironmongery   188
Doorframes and linings   207

7. Windows   **222**

General classification of windows   222
Casement windows and fixed lights   225
Pivot windows   237
Special windows   238
Sliding Sash Windows   244
Glazing to windows   251
Window ironmongery   255
Fixing windows   257
Window boards   260

8. Mouldings and trim   **262**

Timber trim and architrave types   262
Installing architraves   265
Installing skirting   268
Installing dado and picture rails   277

9. Units, fitments, seating and miscellaneous items   **279**

Timber units   279
Retail units and fitments   295
Fixing joinery items   301

10. Casings, false ceilings and panelling   **313**

Casings   313
False ceilings   319
Panelling   324
Cladding   333

11. Temporary construction and site works   **337**

Centring   337
Shoring   340
Formwork   349
Pre-cast concrete moulds   392
Formwork inspection and safety   395
Hoardings and fencing   402

12. Repairs and maintenance   **410**

Maintenance and repair routines   410
Causes of failure in products, processes and materials   411
Inspections and repairs   419
Glazing and painting of doors and windows   433
Multi-skill maintenance   437

Index   **450**

Introduction

This set of two books by Peter Brett is intended to form a complete reference work for carpenters and joiners at all levels. They provide an in-depth knowledge of the basic methods and principles of the craft which are essential to the new entrant and are extended to cover the more specialist topics that may be encountered during their working life.

- *Book 1: Introductory Skills and Job Knowledge* covers the introductory skills and an appreciation of the materials used in the woodworking industry. This is followed up with coverage of the background knowledge that is required to undertake a wide range of practical activities. The final chapters give an overview of the building industry and its controls, which are desirable for understanding and progression.

- *Book 2: Practical Skills for Site and Bench Work* is divided into the specific areas of work which may be undertaken by the carpenter and joiner. The introductory topics in each chapter are covered in depth, supported by bulleted step-by-step text and illustrations, which are intended to guide the reader through the task. This is reinforced by a wider coverage of each topic area to aid understanding and enable the transfer/application of skills to other tasks.

The works are presented in a clear, concise and factual style that is fully integrated with numerous illustrations and photographs. Cross-references and essential key facts are included at the foot of each page, where appropriate, as an aid to full comprehension.

Assessment activities and answers are available on the Web for each chapter to aid learning and enable readers to evaluate their understanding of the knowledge required. See www.nelsonthornes.com/carpentry

These two books combined will prove to be invaluable to both the new entrant undergoing training and qualified craftspeople who are undertaking a particular task for the first time or who wish to extend their knowledge of craft processes and the interrelationship with activities of the building industry as a whole. Additionally, they are ideal background reading for people studying for a career in the building industry or who are already employed at various levels within the industry.

Peter Brett has written numerous books on carpentry and joinery and has extensive experience in the industry and in education and training. He has worked with both the City and Guilds of London Institute and the Construction Industry Training Board in the development and assessment of awards in Carpentry and Joinery.

Carpentry and Joinery
by Peter Brett

Book 1: Introductory Skills and Job Knowledge

1. Timber technology
2. Care and use of hand tools
3. Woodworking joints
4. Interpretation of drawings
5. Portable powered hand tools
6. Basic woodworking machines
7. Site and workshop measurement
8. Applied calculations
9. Applied geometry
10. Building controls
11. Building principles
12. Building procedures

Book 2: Practical Skills for Site and Bench Work

1. Timber ground and upper floor construction
2. Flat and pitched roof construction
3. Partition walls
4. Timber frame construction and timber engineered components
5. Stairs
6. Doors, frames and linings
7. Windows
8. Mouldings and trim
9. Units, fitments, seating and miscellaneous items
10. Casings, false ceilings and panelling
11. Temporary construction and site works
12. Repairs and maintenance

1

Chapter One

Timber ground and upper floor construction

This chapter covers the work of the site carpenter. It is concerned with principles of ground and upper floor construction. It includes the following:

¶ an introduction to floors
¶ timber floor construction
¶ decking, boards and sheets
¶ insulation and finishing.

Floors are primary elements of a building. They form the inner horizontal surface of a room or area.

Types of floor

Floors may be divided into two groups: ground floors and upper floors.

The main functions of a floor are to provide:

■ An acceptable surface for walking, living and working.
■ Acceptable levels of thermal and sound insulation.
■ To support and transfer any loads imposed upon it.
■ Ground floors are also required to prevent moisture penetration and weed growth.

Ground floors

There are two types of ground floor construction: solid and suspended or hollow. A solid ground floor is in direct contact with the ground and a suspended floor spans from wall to wall.

Solid ground floors

These are built up from a number of layers as illustrated in Figure 1.1. Hardcore provides a suitable base for the construction. A thin blinding layer of sand or weak concrete is often used to fill voids or level out the rough edges of the hardcore in order to reduce grout loss or the risk of puncturing the damp-proof membrane (DPM). The DPM is placed

▲ **Figure 1.1** Solid ground floor construction

between the hardcore and concrete slab to prevent dampness rising through the floor. It is essential that that DPM is lapped into the wall's damp-proof course (DPC) in order to prevent any possibility of ground moisture bypassing the DPM. The concrete floor slab (also known as the oversite concrete since it covers the whole site of the building) is not built into the walls but acts as an independent raft transferring its imposed loading directly to the ground and not the foundations.

For industrial use, warehouses and garages etc., the concrete may form the actual floor finish. In other buildings, it is normal to lay a cement and sand screed. This screed levels the slab, takes out any irregularities and provides a smooth surface to receive the final floor finish, e.g. carpet, lino, tiles, woodblocks or strips etc. Alternatively a floating timber floor can be laid over the concrete sub-floor to provide the final finish.

Suspended ground floors

Traditionally these were constructed using timber joists and floorboading, although they are rarely used today. The modern method is to use pre-cast concrete beams, in-filled with concrete blocks, known as beam and block floors.

Timber ground floors. In common with solid ground floors, these still require the hardcore and oversite concrete layers. The dwarf or sleeper walls which support the timber floor construction are built on top of the oversite at about 1.8 m centres using honeycomb bond (a bond that leaves half-brick voids in the wall to allow air circulation). Airbricks must also be included in the outside walls at ground level to enable through ventilation in the under-floor space. This prevents the moisture content of the timber rising too high. (Dry-rot fungi will almost inevitably attack timber in poorly ventilated areas with moisture content above 20%.) The timber floor construction consists of floorboards or sheets, supported by joists that bear on wall plates which in turn spread the load evenly along the sleeper walls. A DPC is placed between the wall plate and sleeper walls to prevent the rise of ground moisture into the timberwork. See Figure 1.2

Beam and block ground floors. The ends of the inverted 'tee' section pre-cast concrete beams are supported on either the inner skin

◄ **Figure 1.2** Timber suspended ground floor

of the cavity wall or by separate sleeper walls. Intermediate sleeper walls may also be required, to provide mid span support or to enable beams to be joined in length. The beams should be evenly spaced to receive the standard size (440 × 215 × 100 mm) concrete infill blocks. These are laid edge to edge and are supported on the beam's protruding bottom edge. See Figure 1.3.

After laying, any gaps between the blocks and beams can be filled by brushing in a sand and cement grout. The actual finished floor surface can be formed using either a sand and cement screed or a timber floor.

◄ **Figure 1.3** Beam and block suspended ground floor

Upper floors

These are also known as suspended floors. Timber construction is mainly used for house construction and concrete for other works.

Timber suspended upper floors. These consist of a number of bridging joists supported at either end by load-bearing walls. The joists are covered on their top by floorboarding or sheeting to provide the floor surface and on their bottom with plasterboard to form the ceiling. The end of the joists can either be supported by building them into the internal leaf of the cavity wall or alternatively, metal joist hangers may be used; these have a flange that locates in the joint of the wall. Where openings in the floor are required for stairs and chimneybreasts etc., the joists must be framed or trimmed around the opening. This entails cutting short a number of bridging joists so that they do not protrude into the opening. These are called trimmed joists. See Figure 1.4.

▶**Figure 1.4** Timber suspended upper floor

Trimmer and trimming joists, which are thicker than the bridging, are framed around the opening to provide support for the ends of the trimmed joists. In order to prevent joists longer than 2 m buckling when under load, they must be strutted at their centre. This strutting has the effect of stiffening the whole floor.

Timber upper floors may be classified into three groups according to their method of construction, see Figure 1.5:

■ *Single floors:* These are the most common type of upper floor for domestic house construction, where common or bridging joists span the shortest dimension of an area from wall to wall.

■ *Double floors:* These are used where the shortest span of the floor is over about 4.8 m or the bridging joists run in the longest dimension. In this case the area can be sub-divided into a number of bays, each of which is not more than 4.8 m long, by binders, which provide intermediate support for the bridging joists.

■ *Triple or framed floors:* These are now obsolete in new construction, as floors where the shortest span is sufficiently long to require them, will almost certainly be constructed using other forms of floor construction. However, triple floors may still be encountered by carpenters and joiners when undertaking maintenance or renovation work. The area is divided up into bays by very heavy, large-sectioned, steel universal beams or concrete beams. These beams provide support for the binders, which in turn provide support for the bridging joists.

Single floor

Double floor

Triple floor

Concrete suspended upper floors. These are supported on the walls or structural framework. They may be formed using either reinforced in situ cast concrete or some form of pre cast units. See Figure 1.6. In both cases, the top surface will require either screeding to receive the floor finish or a timber floor finish and the bottom surface will need plastering or another form of finishing to provide the ceiling

◄ **Figure 1.6** Concrete suspended upper floor

5

surface. In situ concrete floors will require the erection of formwork to cast and support the wet concrete. In general, concrete floors are more sound resistant to airborne noise and have better fire-resistant properties than do timber floors.

Timber floor construction

Joists

These are of a series of parallel timber beams used to span the gap between walls and directly support a floor surface and the ceiling surface below. See Figure 1.7.

Boarding fixed to joists to form floor surface

Noggins between joists to support plasterboard around perimeter and at joints

Plasterboard ceiling fixed to underside of joists

Joists spanning from wall to wall

▶**Figure 1.7** Joists

Sectional size of joists

The sectional size of a joist (breadth and depth) depends on its span, spacing, weight or loading placed upon it and the strength class of the timber used. For new work the sectional size will always be specified. The Building Regulations Approved Document (AD) Part A contains tables of suitable sectional joist size for use in different situations, see Table 1.1 for an example.

The approximate depth of 50-mm breadth joists spaced at 400-mm centres can be found by using the following rule-of-thumb formula:

Depth of joist = {(Span of joist in millimetres)/20} + 20

Example

for joists spanning 4000 mm
Depth of joist = (4000/20) + 20 = 220 mm

The nearest commercial size of timber to use would be 50 × 225 mm or 50 × 220 mm regularised. These are suitable for normal domestic use, although on the generous side, as by reference to Table 1.1 it

Table 1.1 Maximum clear span of floor joists (m)

| Breadth × Depth | Spacing (centre to centre) | | | Based on using C16 grade softwood with a dead load excluding weight of joist between 0.25 and 0.5 kN/m² |

Size of joist (mm)	400 mm	450 mm	600 mm	Example
38 × 97	1.72	1.56	1.21	To determine size of joists, spaced at 600 mm
38 × 122	2.37	2.22	1.76	centres and having a clear span of 3.75 m:
38 × 147	2.85	2.71	2.33	
38 × 170	3.28	3.10	2.69	1. Look down the 600 mm column to get the
38 × 195	3.72	3.52	3.06	nearest value above 3.75 m
38 × 220	4.16	3.93	3.42	
				2. This is 3.90 m.
50 × 97	1.98	1.87	1.54	Move to the left hand column to get the
50 × 122	2.60	2.50	2.19	required joist size.
50 × 147	3.13	3.01	2.69	
50 × 170	3.61	3.47	3.08	3. 50 × 220 mm joists are required in this case.
50 × 195	4.13	3.97	3.50	
50 × 220	4.64	4.47	3.90	4. As an alternative 75 × 195 mm joists could be
				used, but would not be economical
75 × 122	2.97	2.86	2.60	
75 × 147	3.56	3.43	3.13	
75 × 170	4.11	3.96	3.61	
75 × 195	4.68	4.52	4.13	
75 × 220	5.11	4.97	4.64	

can be seen that 50 × 195 mm joists could be used for spans up to 4130 mm.

Span of joists

- *Clear span* is the distance between joist supports.
- *Effective span* is the distance between the centres of the joist bearings.

▲ **Figure 1.8** Span of joists

The bearing itself is the length of the end of a joist that rests on the support. The overall length of a joist is its clear span plus the length of its end bearings, e.g. a joist with a 3400-mm clear span and 100-mm end bearings will have an effective span of 3500 mm and an overall joist length of 3600 mm. See Figure 1.8.

Joists are commonly laid out to span the shortest distance between the supporting walls of a room or other area. This keeps to a minimum the size of joist required. Once the depth of the joist is determined for the longest span it is normal practice to keep all other joists the same. The shorter span joists will be oversize, but all joist coverings and ceiling surfaces will be level.

Spacing of joists

Joist spacing is the distance between the centres of adjacent joists. Commonly called joist centres or c/c (centre to centre) see Figure 1.9. They range between 400 to 600 mm depending on the joist covering

◄ **Figure 1.9** Spacing of joists

Table 1.2 Guide to maximum span of decking (joist covering materials)

Decking material	Finished thickness	Maximum span
		Span = joist spacing
Softwood planed, tongued and grooved P, T & G	16 mm	450 mm
Softwood planed, tongued and grooved P, T & G	19 mm	600 mm
Flooring grade chipboard	18 mm	450 mm
Flooring grade chipboard	22 mm	600 mm
Flooring grade plywood	16 mm	400 mm
Flooring grade plywood	19 mm	600 mm

material. Joists should be spaced to accommodate surface dimensions of their covering material; see Table 1.2 for maximum joist spacings for a range of joist covering materials.

End joists adjacent to walls should be kept 50 mm away from the wall surface, in order to allow an air circulation and prevent dampness being transferred from wall to joist and thus reduce the risk of rot. In addition this gap helps reduce the transmission of noise at party walls.

Unventilated and damp timber is open to an attack of rot (dry rot or wet rot), which leads to a loss of strength and possible collapse. For detailed information on timber rots, their prevention and remedial treatment see Book 1.

To determine the number of joists required and their centres for a particular area the following procedure, shown in Figure 1.10, can be used:

1. Measure the distance between adjacent walls say 3150 mm. The first and last joist would be positioned 50 mm away from the walls. The centres of 50 mm breadth joists would be 75 mm away from the wall. The total distance between end joists centres would be 3000 mm.

2. Divide distance between end joist centres by specified joist spacing say 400 mm. This gives the number of spaces between joists. Where a whole number is not achieved, round up to the nearest whole number above. There will always be one more joist than the number of spaces so remember to add one to this figure to determine the number of joists.

Where T & G boarding is used as a floor covering the joist centres may be spaced out evenly, i.e. divide the distance between end joist centres by the number of spaces.

Where sheet material is used as a joist covering to form a floor, ceiling or roof surface, the joist centres are normally maintained at a 400-mm or 600-mm modular spacing to coincide with sheet sizes. This would leave an undersized space between the last two joists.

▲ **Figure 1.10** Determining the number of joists required

Joist sections

The strength of a joist varies in direct proportion to changes in its breadth and in proportion to the square of its depth. For example, doubling the breadth of a section doubles its strength (see Figure 1.11):

> A 100 mm × 100 mm joist has double the strength of a 50 mm × 100 mm joist.

Doubling the depth of a section increases its strength by four times:

> A 50 mm × 200 mm joist has four times the strength of a 50 mm × 100 mm joist.

Less material is required for the same strength when the greatest sectional dimension is placed vertically rather than horizontally:

> A 100 mm × 100 mm joist has the same sectional area as a 50 mm × 200 mm joist, but the deeper joist would be twice as strong.

▲ **Figure 1.11** Joist sections

Therefore joists are normally placed so that the depth is the greatest sectional dimension. Joists of the same sectional size and span would clearly have different strengths if their breadths and depths were reversed. Those with the smaller depth would sag under load, possibly leading to structural collapse.

When stating the sectional size of a joist the first measurement given is the breadth and the second the depth.

Regularised timber (Termed machined timber in European Regulations). This is preferred for joists as all timber will be a consistent depth. This aids floor levelling and ensures a flat fixing surface for joist coverings and ceilings. Joists are regularised to a consistent depth by re-sawing or planing one or both edges. A reduction in size of 3 mm must be allowed for timber up to 150 mm in depth and 5 mm over this depth. See Figure 1.12. Thus, 50 mm × 150 mm may be regularised to 50 mm × 147 mm and 50 mm × 200 mm may be regularised to 50 mm × 195 mm.

▲ **Figure 1.12** Regularised (machined) joist

Positioning of joist defects. Any joists which are not straight should be positioned with their camber or crown (curved edge) upwards. When loaded these joists will tend to straighten out rather than sag further if laid the other way. Joists with edge knots should be positioned with the knots on their upper edge. When loaded the knots will be held in position as the joists sag, rather than fall out, weaken the joist, and possibly lead to structural collapse if laid the other way. See Figure 1.13.

Camber upwards joist will straighten under load

Would sag further under load if camber downwards

Edge knot

Edge knots held in place under load

Edge knot

Edge knots fall out under load leading to collapse

▶**Figure 1.13** Positioning of joists with camber or edge knots

Joist supports and restraints

Joist supports

The ends of joists may be supported:

- by building in
- on hangers
- on wall plates
- on binders.

Building in the inner leaf of a cavity wall. The minimum bearing in a wall is normally 90 mm, see Figure 1.14. Shorter bearings do not tie in the wall sufficiently and can lead to a crushing of the joist end possibly leading to collapse. A steel bearing bar may be

TIMBER FLOOR

Floor boards

End built into wall

Timber joist

▲ **Figure 1.14** Building in a joist to 90 mm minimum

incorporated into the mortar joint where lightweight blocks are used to reduce the risk of the blocks crumbling. The ends of the joists are often splayed but they must not project into the cavity where they could possibly catch mortar droppings during building, leading to dampness in the joist and rotting. The ends of joists that are in contact with the external wall should be treated with a timber preservative to protect them from dampness and subsequent rot.

Hangers. Ends of joists may be supported on galvanised steel joist hangers, which are themselves built into or bear on a wall (see Figure 1.15). Double hangers are available which saddle internal walls to provide a support for joists on both sides. An advantage of this method is that the joist can be positioned independently of the building process. Hangers are useful when forming extensions as they are simply inserted into a raked-out mortar joint. The bottom edge or bearing surface of the hanger must be recessed into the joist to prevent the hangers obstructing any ceiling covering. Joists should be secured into the hanger using 32-mm galvanised clout nails in each hole provided.

▲ **Figure 1.15** Use of a joist hanger

In modern construction, the preferred method of supporting the ends of upper floor joists in external walls is to use joist hangers, as any subsequent timber shrinkage or warping may result in gaps in the wall around a built-in end leading to a reduction in both sound and thermal insulation.

Wall plates. Ends of joists may be supported by a wall plate bedded on the top of a wall. This is normally used for ground floor construction and over internal load-bearing partitions for upper floors (see Figure 1.16). The minimum bearing recommended on a wall plate and thus the minimum width is 75 mm, although 100 mm wide wall plates are normally used for ground floor sleeper walls. Often joists from either side meet over a wall plate; it is usual to nail them together side by side, both overlapping the wall plate by about 300 mm.

▲ **Figure 1.16** Joist supported on a wall plate

The use of wall plates provides a means of securing joists by skew nailing with 75 mm or 100 mm wire nails. In addition, wall plates also spread the loading of a joisted surface evenly over a wide area rather than a point load. Cambered joists may be straightened over a wall plate by partly sawing through and nailing down. Wall plates are not suitable for use in external walls of upper floor construction. This is due to shrinkage movement and the likelihood of rot.

The carpenter uses halving joints as shown in Figure 1.17 to join wall plates at corners and where required in length. The plates are then normally bedded and levelled in position by the bricklayer using bedding mortar.

▶ **Figure 1.17** Wall plate joint

Intermediate support to joists

Binders. These are introduced into a floor structure in order to provide an intermediate support for large span joists used in double and triple floors. These binders may be of:

■ steel (when they are known as a universal beam (UB);

■ timber (either of solid section, glue laminated section (glulam) or a plywood box beam);

■ or concrete.

Depending on the space available, binders may be positioned below the joists, or accommodated partly within the joist depth projecting above or below as required. See Figure 1.18. Where joists are fitted to a steel universal beam, a plywood template may be cut to speed the marking out of the joist end and ensure a consistent, accurate fit. (See Figure 1.19).

▲ **Figure 1.18** Using binders

Fire protection of steel binders. Where universal beams are used in floor construction they must be protected as a fire precaution. This protection is required in order to prevent early structural collapse resulting from the fact that the strength of steel rapidly decreases at temperatures above 300ºC. Various methods of protection include:

■ a sprayed coat of non-combustible insulating material;

■ a solid casing of concrete;

- encasing in expanded metal lathing and plastered;
- encasing with plasterboard or other non-combustible material.

Figure 1.20 shows a typical method of encasing a steel beam using plasterboard and a skim coat of finishing plaster. The plasterboard has been fixed to timber cradling members nailed to each joist.

Joist restraints

Restraint straps. With modern lightweight structures, walls and joisted areas require positive tying together for strength, to ensure wall stability in windy conditions. This is a requirement of the Building Regulations. Galvanised mild steel restraint straps, also known as lateral restraint straps, must be used at not more than 2-m intervals and carried over at least three joists. Noggins should be used under the straps where the joists run parallel to the wall; these should be a minimum of 38 mm thick and extend at least half the depth of the joist. For example with 200-mm deep joists at least 100 mm deep noggins are required. See Figure 1.21.

Strutting. Where deep joists exceed a span of 2.5 m, they tend to buckle and/or sag under loads unless restrained by strutting. Joists spanning between 2.5 and 4.5 m should be restrained by strutting at their mid span; larger span joists over 4.5 m will require two rows of strutting at one third and two thirds of their span. Strutting is a requirement of the Building Regulations.

▲ **Figure 1.19** Marking a joist to fit a beam

▲ **Figure 1.20** Plasterboard and cradling to a universal beam

◀ **Figure 1.21** Restraint strap

▲ **Figure 1.22** Strutting to joists

There are three main types of strutting in use see Figure 1.22:

■ *Solid strutting* is quick to install but is considered inferior, as it tends to loosen and become ineffective when joists shrink. It should be at least 38 mm thick and extend to at least $\frac{3}{4}$ of the joist's depth, e.g. a 200-mm deep joist will require at least 150 mm deep struts.

■ *Herringbone timber strutting* is considered the most effective as it actually tightens when joists shrink. However, it takes longer to install and thus is more expensive in terms of labour costs. The minimum sectional size is 38 mm × 38 mm. Timber herringbone struts can only be used where the spacing between the joists is less than three times the depth of the joist, e.g. for 150-mm deep joists herringbone struts are suitable up to 450 mm spacing, but for 200-mm deep joists a spacing of up to 600 mm is permitted.

■ *Galvanised steel herringbone strutting* is quick to install as no cutting is required, but has a disadvantage in that the depth and centres of the joist must be specified when ordering; different depths and spacings will require different sized struts.

Herringbone timber struts can be marked out using the following procedure shown in Figure 1.23

1. Mark across the joists the centre line of the strutting.

2. Mark a second line across the joist so that the distance between the lines is 10 mm less than the depth of the joist.

3. Place the length of strutting on top of the joists as shown and mark underneath against the joists at A and B.

4. Cut two struts to these marks. If all joists are spaced evenly, all the strutting will be the same size and can be cut using the first one as a template. If not, each set of struts will have to be marked individually.

5. Fix struts on either side of the centre line using wire nails, one in the top and bottom of each strut and one through the centre.

▶**Figure 1.23** Marking out strutting

The gaps between the end joists will require packing and wedging to complete the system. See Figure 1.24.

▶**Figure 1.24** Packing and wedging

Whichever method of strutting is used, care should be taken to ensure that they are clear of the tops and bottoms of the joists, otherwise they may subsequently distort the joist covering or ceiling surface.

Again, whichever method of strutting is used, the gaps between the end joists and the walls will require packing and wedging to complete the system. Care must be taken not to over tighten the folding wedges as it is possible to dislodge the blockwork.

Jointing and positioning joists

Trimming

Where openings are required in joisted areas or where projections occur in supporting walls, joists must be framed or trimmed around them (see Figure 1.25). Members used for trimming each have their own function and are named accordingly:

■ *Bridging joist:* a joist spanning from support to support, also known as a common joist.

■ *Trimmed joist:* a bridging joist that has been cut short (trimmed) to form an opening in the floor.

■ *Trimmer joist:* a joist placed at right angles to the bridging joist in order to support the cut ends of the trimmed joists.

■ *Trimming joist:* a joist with a span the same as the bridging joist but supporting the end of a trimmer joist.

As both the trimmer and the trimming joists take a greater load, they are usually 25 mm thicker in breadth than the bridging joists. For example, use 75 mm × 200 mm trimmer and trimming joists with 50 mm × 200 mm bridging joists.

Trimming joints. Traditionally, **tusk mortise and tenon joints** were used between the trimmer and trimming joists (Figure 1.26), while housing joints were used between the trimmed joists and the trimmer. Once wedged, the tusk mortise and tenon joint requires no further fixing. The housing joints will require securing with 100-mm wire nails. The proportions of these joints must be followed. They are based on the fact that there are neutral stress areas in joists (Figure 1.27). If any cutting is restricted mainly to this area then the reduction in the joist strength will be kept to a minimum.

▲ **Figure 1.25** Trimming openings

'dovetail joint'

wedge to pull joint up

tusk mortise and tenon

bevelled housing

$\frac{1}{2}$D

$\frac{1}{3}$W

width W

depth D

$\frac{1}{3}$D

$\frac{1}{6}$D

$\frac{1}{4}$D

$\frac{1}{4}$D

$\frac{1}{4}$W   150 mm

square housing

$\frac{1}{2}$D

$\frac{1}{4}$W

▲ **Figure 1.26** Traditional trimming joists

Joints cut in shaded areas
will cause the minimum
reduction in strength

▲ **Figure 1.27** Neutral stress areas

▲ **Figure 1.28** Trimming using a joist hanger

Joist hangers (Figure 1.28), which are a quicker, modern alternative to the traditional joints, are now used almost exclusively. As these hangers are made from thin galvanised steel, they do not require recessing in as do the thicker wall hangers but they must be securely nailed in each hole provided with 32-mm galvanised clout nails.

To provide sufficient headroom for quarter-turn stairs, an L-shaped stairwell opening is often required. Figure 1.29 shows how the landing is formed using **cantilever joists**, which must be carried through the wall and across the adjacent room. Solid strutting should be fixed diagonally across the landing for strength. Where the joists meet at the external angle a dovetail joint can be used.

▲ **Figure 1.29** L-shaped stairwell

Positioning joists

Vertical position of joists. This will normally have been established by the bricklayers. They will have built up to the required brickwork course for the underside of built-in floor joists or bedded a wall plate on at this position. When joist hangers are being used the brickwork is built up to the required course to take the lug of the hanger. This will be near the top edge of the joist. See Figure 1.30. Alternatively where straight lugged hangers, rather than the turned down restraint hanger, are used the wall may be built beyond the floor level and the bed joint is raked out to receive the hanger.

Laying out joists. Joists are first cut to length, see Figure 1.31. Ground floor joists should be cut between the walls to give a gap of about 20 mm at both ends. Upper floor joists that are built in should be cut with a splayed end to avoid any possibility of the

▲ **Figure 1.30** Vertical position of joists

20 mm gap

Ends of ground floor joist kept clear of wall

Splayed end cut

✔ Ends of upper floor joists 'built in' kept clear of cavity

✘

✔ Ends of upper floor joists cut tight to hangers

✘ Too tight may result in wall movement

✘ Too short gives a reduced bearing and the hanger may be displaced

Crushing to end of joist due to lack of bearing

▲ **Figure 1.31** Cutting joists to length

end protruding into the cavity. Joists in hangers must be cut to the exact length between hangers to keep the hangers tight against the wall. Take care not to cut them too long, as forcing them in can easily push the freshly laid walls out of position.

The outside joists are placed in position first, leaving a 50-mm gap between them and the wall. The other joists are then spaced out to the required centres in the remaining area.

However, when openings for fireplaces and stairwells, etc., are required in the joisted area the layout of the joists is governed by these openings. The trimming and outside bridging joists are the first to be positioned, once again leaving a 50-mm gap between all joists and the walls. The other joists are then spaced out to the required centres in the remaining areas.

Timber near sources of heat

The Building Regulations restrict the positioning and trimming of timber near sources of heat. They state that the construction around heat-producing appliances should minimise the risk of the building catching fire. In order to comply with this requirement the use of combustible material in the vicinity of a fireplace, chimney or flue is restricted, see Figure 1.32.

■ No timber is to be built into the flue, or be within 200 mm of the flue lining, or be nearer than 40 mm to the outer surface of a

Flooring, skirting and other trim don't require 40 mm separation

Metal fixings at least 50 mm

Joist

Minimum of 40 mm

200 mm for structural timber

Flue wall

Flue lining

150 mm minimum

125 mm minimum concrete

Dovetail strip

500 mm

125 mm concrete

50 mm airspace

250 mm separation

Calcium silicate board to act as hearth formwork

Trimming joist

▶ **Figure 1.32** Timber construction near a source of heat

chimney or fireplace recess. (A 50-mm airspace between any joist and wall is standard practice.) The exceptions to this requirement are: flooring; skirting; architrave; mantelshelf and other trim. However any metal fixings associated with these must be at least 50 mm from the flue.

■ There must be a 125-mm minimum thickness solid non-combustible hearth, which extends at least 500 mm in front of the fireplace recess and 150 mm on both sides.

■ Combustible material used underneath the hearth is to be separated from the hearth by airspace of at least 50 mm or be at least 250 mm from the hearth's top surface. Combustible material supporting the edge of the hearth is permitted.

Ground floors. The layout of joists around a fireplace is illustrated in Figures 1.33 and 34. The ends of the joists in front of the fireplace are supported by a wall plate on a fender wall. The fire hearth is framed out with lengths of joist which sit on the fender wall. This framing performs two functions:

■ It acts as formwork when the fire hearth is concreted.

■ It supports the ends of the floor boards around the hearth.

Upper floors. Fireplace openings in upper floors are virtually obsolete. However, they are covered in this chapter because they may be met in maintenance or rehabilitation work. Figure 1.35 shows the layout of joists for a single floor with a fireplace opening. The dovetail strips shown in the figure are bedded into the top of the concrete hearth to provide a fixing for the ends of the floorboards. Figure 1.36 illustrates a method of trimming joists around a projecting flue, from a ground floor fireplace or other heat-producing appliance.

Double joists are required where blockwork partition walls are to be built on a joisted area. These double joists are either nailed or bolted together (depending on the specification) and positioned under the intended wall to take the additional load as illustrated in Figure 1.37. When laying out a joisted area, double joists are positioned along with the trimming and end bridging joists, before the other joists are spaced.

Change of joist direction. In order to span the shortest direction it is sometimes necessary to change the direction of joists in a particular joisted area. This is done over a load-bearing wall where the bridging joists from one area are allowed to overhang by 50 mm; the end-bridging joist in the other area is simply nailed to the overhanging ends (see Figure 1.38).

Levelling joists

After positioning, the joists should be checked for line and level. End joists are set using a spirit level; intermediate joists are lined through with a straightedge and spirit level (see Figure 1.39). Where the brickwork or blockwork course has been finished to a level line by the bricklayer or where wall plates are used and have

Plan

▲ **Figure 1.33** Layout of joists around floor fireplace (plan view)

Section

▲ **Figure 1.34** Section view through hollow ground floor with fireplace

▲ **Figure 1.35** Layout of joists around upper floor fireplace

▶**Figure 1.36** Trimming around a projecting flue

been accurately bedded level, and where regularised joists have been used, the bearings of the joist should not require any adjustment to bring them into line and level.

However, minor adjustments may be required as shown in Figure 1.40. Joists may be housed into or packed off wall plates. Where packings are required for built-in joists, these should be slate or other durable

▲ **Figure 1.37** Double joists required to support blockwork above

▲ **Figure 1.38** Typical joist layout showing change of direction

▲ **Figure 1.39** Levelling joists

material. Do not use timber packings as these may shrink and work loose and in any case are susceptible to rot. Joists may be recessed, to lower their bearing, providing the reduced joist depth is still sufficient for its span.

Figure 1.41 shows temporary battens, which can be nailed across the top of the joists to ensure that their spacing remains constant before and during their building in. Joists fixed to wall plates can be skew nailed to them using 75-mm or 100-mm wire nails.

▲ **Figure 1.40** Packing, housing or recessing joists to maintain level

▲ **Figure 1.41** Use of temporary batten to secure joists before building in

Notching and drilling joists for services

The position of notches for pipes and holes for cables in joists should have been determined at the building design stage and indicated on the drawings. Both reduce the joist strength. Notches and holes in joists should therefore be kept to a minimum and conform to the following (see Figure 1.42):

■ Notches on the joist's top edge of up to one eighth (0.125) of the joist's depth located between, 7% and 25% (0.07 and 0.25) of the span from either support are permissible.

■ Holes of up to one quarter (0.25) of the joist's depth drilled on the neutral stress line (centre line) and located between 25% and 40% (0.25 and 0.4) of the span from either support are permissible. Adjacent holes should be separated by at least three times their diameter measured centre to centre.

Example

The position and sizes for notches and holes in a 200-mm depth joist spanning 4000 mm are:
• *Notches* between 280 mm and 1000 mm in from each end of the joist and up to 25 mm deep.
• *Holes* between 1000 mm and 1600 mm in from each end of the joist and up to 50 mm diameter.

Excessive notching and drilling of holes even within the permissible limits will seriously weaken the joist and may lead to structural failure.

▲ **Figure 1.42** Positions for notches and holes

Floor covering and finishing

Floor decking is also variously termed as flooring, floorboarding or joist covering. The main materials in common use for this are illustrated in Figure 1.43:

■ Timber floorboards mainly PTG (planed, tongued and grooved) but square-edged boarding may be found in older buildings.

■ Flooring grade particleboard, mainly tongued and grooved or square-edged chipboard, but OSB (orientated strand board) is also used.

■ Flooring grade plywood either tongued and grooved or square-edged.

Boards   Square edged

tongued and grooved

Boards

Chipboard or OSB   Plywood

◀ **Figure 1.43** Joist coverings

The laying of floor decking is normally carried out after plumbing and electrical under-floor carcassing work and preferably also after the window glazing and roof tiling is complete so that it is not exposed to the weather. Where sheet material is used in kitchens and bathrooms, a water-resisting grade should be used.

Timber flooring boards

Hardwood strip flooring. This is used where the floor is to be a decorative feature. It is made from narrow strips of tongued-and-grooved hardwood. These are usually random lengths up to 1.8 m long and 50 mm to 100 mm in width. The heading joints are also usually tongued and grooved. Figure 1.44 shows a section through a piece of hardwood strip floorboard. It has a splayed-and-stepped tongue to ease secret nailing and also to reduce any likelihood of the tongue splitting. The purpose of the recessed back is to ensure positive contact between the board and joist.

splayed-and-stepped tongue

recessed back

▲ **Figure 1.44** Hardwood strip floorboards

It is essential that the newly laid hardwood strip flooring is protected from possible damage by completely covering it with building paper or a polythene sheet, which should remain in position until the building is ready for occupation.

Softwood flooring. This usually consists of ex 25 mm × 150 mm tongued and grooved boarding. A standard floorboard section has the tongue and groove offset away from the board's face. This identifies the upper face and also provides an increased wearing surface before exposing the tongue and groove. Boards can be fixed either by floor brads nailed through the surface of the boards and punched in, or by lost-head nails secret-fixed through the tongue as shown in Figure 1.45. Nails should be approximately $2\frac{1}{2}$ times the thickness of the floorboard in length.

Floor brad nailed through face and punched in   Lost heads used to secret nail through tongue

▲ **Figure 1.45** Fixing softwood flooring

Square heading

Splayed heading

▲ **Figure 1.46** Heading joints for softwood flooring

Heading joints. Figure 1.46 shows square-butted or splayed heading joints, which are introduced as required to utilise off-cuts of board and avoid wastage. Splayed heading joints are preferred, as there is less risk of the board end splitting and do show a wide gap if shrinkage takes place. Heading joints should be staggered evenly throughout the floor for strength; these should never be placed next to each other, as the joists and covering would not be tied together properly. See Figure 1.47.

Laying timber flooring boards

Surface fixing. Boards are laid at right angles to the joists. The first board should be fixed at least 10 mm away from the outside wall. This gap, which is later covered by the skirting, helps to prevent dampness being absorbed through direct wall contact. In addition, the gap also allows the covering material to expand without either causing pressure on the wall or a bulging of the floor surface. The remainder of the boards are laid four to six at a time, cramped up with floorboard cramps (see Figure 1.48) and surface nailed to the joists. The final nailing of a floor is often termed 'bumping' which should be followed by punching the nail head just below the surface. Boards up to 100 mm in width require one nail to each joist while boards over this require two nails.

▲ **Figure 1.47** Positioning of heading joints

▶**Figure 1.48** Use of floorboard cramp

Tightening floorboards. Figure 1.49 shows two alternatives to the use of floorboard cramps: folding or wedging, although neither of these is as quick or efficient. Folding a floor entails fixing two boards spaced apart 10 mm less than the width of five boards. The five boards can then be placed with their tongues and grooves engaged. A short board is laid across the centre and 'jumped on' to press the boards in position. This process is then repeated across the rest of the floor. Alternatively, the boards may be cramped, four to six at a time using dogs and wedges. Whichever method is used the final board will invariably require cutting to width; rip it about 10 mm narrower than the gap and lever off the wall to give a tight joint before fixing.

Secret fixing. Secret-fixed boards must be laid and tightened individually and cramping is therefore not practical. Figure 1.50 shows how they may be tightened by levering them forward with a firmer chisel driven into the top of the joist, or with the aid of a floorboard nailer. This tightens the boards and drives the nail when the plunger is struck with a hard mallet. Secret fixing is normally only used on high-class work or hardwood flooring, as the increased laying time makes it considerably more expensive.

Services

Where services such as water and gas pipes or electric cables are run within the floor, there is a danger of driving nails into them. They should be marked on saying in chalk or pencil 'PIPES NO FIXING', so that on nailing the danger area is kept clear. The floorboards over services can be fixed with recessed cups and screws to permit easy removal for subsequent access, and also to provide easy recognition of location, see Figure 1.51.

▲ **Figure 1.49** Tightening floorboards: folding and wedging

▲ **Figure 1.50** Tightening secret-fixed boards

▲ **Figure 1.51** Marking position of services

Access traps. These may be required in a floor over areas where water stopcocks or electrical junction boxes are located, see Figure 1.52. Again these can be fixed with recessed cups and screws to permit easy removal.

access trap to be screwed down to bearers

25 mm x 50 mm battens nailed to boards to form access trap

25 mm x 50 mm bearer screwed to joist

50 mm x 100 mm joists

▶**Figure 1.52** Providing access to services

Fireplace openings. Figure 1.53 shows how margin strips are used to finish the ends of the floor boarding around a fire hearth. The actual arrangement will vary depending on whether the boards run into or across the hearth. In either case the layout of boarding should be planned before any fixing takes place, to ensure an even margin around the hearth.

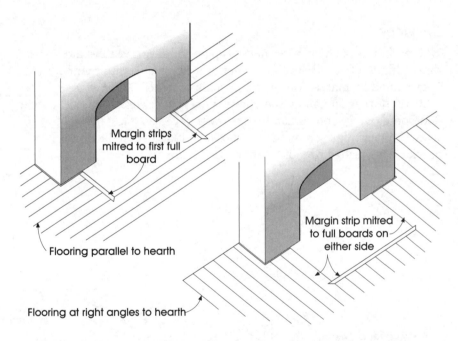

Margin strips mitred to first full board

Flooring parallel to hearth

Margin strip mitred to full boards on either side

Flooring at right angles to hearth

▲ **Figure 1.53** Layout and finish to flooring around hearth

Sheet floor decking

This is now being increasingly used for domestic flooring. Flooring grade chipboard is available with square edges in 1220 mm × 2440 mm sheets, and with tongued and grooved edges in 600 mm × 2440 mm sheets. Square-edged sheets are normally laid with their long edges over a joist. Noggins must be fixed between the joists to support the short ends. Tongued and grooved sheets are usually laid with their long edges at right angles to the joists and their short edges joining over the joist. Both types require noggins between the joists where the sheet abuts a wall. See Figure 1.54. Joists should be spaced to accommodate the dimensions of the sheet flooring.

▲ **Figure 1.54** Layout of chipboard floors

Fixing sheet flooring

Sheets are laid staggered and fixed at 200 mm to 300 mm centres with 50 mm lost-head nails or, for additional strength annular ring shanked or serrated nails. A gap of 10 mm must be left along each wall to allow for expansion and prevent absorption of dampness from the wall. Manufacturers of chipboard often recommend the gluing of the tongues and grooves with a PVA adhesive to prevent joint movement and to stiffen the floor. See Figure 1.55.

For protection it is recommended that the floor be covered with building paper after laying and that this is left in position until the building is occupied.

Plywood and OSB sheet floor decking are laid using the same procedures as chipboard flooring.

▲ **Figure 1.55** Gluing joints in chipboard flooring

Insulation, preservation and finishing

Insulation of floors

Thermal insulation

Suspended timber ground floors should be insulated to prevent the cool ventilated under floor space drawing the warm air from the room. This can be achieved be placing insulating material between the joists, see Figure 1.56. Timber bearers nailed to the bottom edges of the joists can be used to support rigid panels. Alternatively a plastic meshing can by laid over and down between the joists, forming a tray on which a blanket insulation can be supported. In addition the

Timber bearers nailed to
joists support rigid
insulation

Plastic meshing laid over
joists forms a tray to
support blanket insulation

▶**Figure 1.56** Thermal insulation
to suspended timber ground floors

insulation can be further improved by laying sheets of fibre insulation
board over the joists before fixing the floor decking.

Suspended timber upper floors do not normally require added
thermal insulation except where they separate a heated room from an
unheated room or space below, see Figure 1.57. In these circumstances
the space between the joists can be filled with either fibreglass or
mineral wool insulation resting on the ceiling.

Insulation between joists
supported on ceiling

▶**Figure 1.57** Thermal insulation
to suspended timber upper floors

Sound insulation

The need for sound insulation in domestic floors is fairly limited. Where
required in upper floors to separate say a children's noisy bedroom from
the lounge below, the details shown in Figure 1.58 can be used:

■ Filling the space between the joists with a sound insulating slag
wool pugging or blanket insulation will reduce the passage of
air-borne sound.

Airborne

Resilient strip to seal gap

Wool pugging or
blanket insulation
between joists

End space insulated

Impact

Resilient strip under skirting

Flooring fixed to battens only

Loose laid battens

Resilient quilt

Quilt turned up behind skirting

Airborne and Impact

◀ **Figure 1.58** Sound insulation to
suspended timber upper floors

- A floating floor will insulate against impact or structural-borne sound. This uses a resilient quilt to separate the floor decking from the floor structure. To be effective the battens that support the flooring should be loose laid over the joists and not fixed to them.

- A combination of the previous two methods can be used to reduce both air-borne and structural-borne sound.

The passage of sound will find the point of least resistance; therefore do not forget to insulate the 50 mm gap between the outside joists and the wall. In addition a resilient seal should be used at the skirting to floor intersection.

Preservative treatment of joists

It is recommended that all timber used for structural purposes is preservative treated before use. Any preservative treated timber cut to size on site will require re-treatment on the freshly cut edges/ends. Applying two brush flood coats of preservative will achieve this. Timber preservatives prevent rot by poisoning the food supply on which fungi feed and grow.

Timber finishes to solid floors

There are three main methods, which can be used to provide a timber finish to solid floors:

- Timber floor boarding or sheets fixed to timber battens.
- Sheet floor panels laid on rigid insulation.
- Wood blocks bonded directly to the concrete.

Timber floor boarding or sheets fixed to timber battens

The timber bearers or battens may be bedded in, clipped to, fixed directly to, or left floating on the solid floor, see Figure 1.59. These bearers, which are spaced at about 400 or 600 mm centres, act as floor joists and thus provide a means of fixing the floorboards.

- *Dovetail-shaped bearers* are used for bedding in. Small mounds of screed or concrete are deposited along the bearer line into which the bearers are tapped down to a level position. When set, further screed or concrete can be laid using the bearers as a guide for levelling off.

- *Floor clips* are made from folded galvanised mild steel and are available in either a standard pattern or with a separating resilient pad, known as an acoustic pattern (used to increase the impact sound insulation of solid upper floors). These are pushed into freshly laid screed or concrete, along the bearer lines at about 600 mm centres. When set, the lugs of the clips are opened up to form a cradle in which the bearers are located. Special improved nails are use to secure the bearers to the clips.

- *Directly fixed bearers* are normally shot fixed to the concrete at about 600 mm centres using a cartridge operated fixing tool.

▲ **Figure 1.59** Timber floor finish

- *Floating bearers* are laid unfixed on a layer of resilient insulating material. When the floor boarding is fixed to the bearers the whole floor floats on the insulation and is held down by its own weight and the skirting around the perimeter of the room.

Insulation. In order to provide thermal insulation the space between the bearers can be filled with fibreglass or mineral wool. Alternatively a layer of insulating fibreboard can be used between the bearers and the floor boarding.

Vapour barrier. A polythene vapour barrier should be laid over the bearers or fibreboard before fixing the floor boarding in order to prevent condensation forming in the floor space. The vapour barrier should be turned up the walls at its edges and any joints overlapped and sealed with a waterproof tape.

Sheet floor panels laid on rigid insulation

Tongued and grooved plywood or chipboard flooring panels are laid directly and floating on a rigid layer of insulation, normally polystyrene. See Figure 1.60. The joints of the sheeting should be glued as they are laid, to provide an insulating floor finish that is held down by its own weight and the skirting around the perimeter of the room. Composite flooring panels are also available for this purpose, with the insulation being pre-bonded to the back.

▲ **Figure 1.60** Sheet flooring on rigid insulation

Wood blocks bonded directly to the concrete

This method is mostly used for laying wood block or parquet floor finishes using a bituminous mastic compound as the adhesive. See Figure 1.61. An expansion gap around the perimeter of the floor is required; this should be filled with either polystyrene or cork so that the blocks are allowed an amount of moisture movement but are prevented from 'creeping' and closing the gap.

Hardwood strip margin

Wood blocks laid in a
herring-bone pattern

Resilient strip to allow
block moisture
movement later
concealed by skirting

Bituminous compound used
to bond blocks to concrete

Width

Length
= 3x width

Bottom edge of hardwood block
dovetailed to key into bitumin

▲ **Figure 1.61** Wood block floor

2

Chapter Two

Flat and pitched roof construction

This chapter covers the work of the site carpenter. It is concerned with principles of flat and pitched roof construction. It includes the following:

- ❡ an introduction to roof types
- ❡ timber flat roof construction
- ❡ traditional pitched roof construction
- ❡ pitched roof construction using trussed rafters
- ❡ verges and eaves
- ❡ dormer windows and ornamental roofs
- ❡ roof erection
- ❡ prefabricated rafters.

Roofs are part of the external envelope that spans the building at high level and has structural, weathering and insulation functions.

Roof types

Roofs are classified according to their pitch, shape and also type of structure.

Roof pitch

The pitch or slope of a roof surface to the horizontal may be expressed either in degrees or as a ratio of the rise to the span, see Figure 2.1. For example: a flat roof with a slope of 1:40 will rise 1 unit for every 40 units run, that is a rise of 25 mm for every metre run; a one third pitch roof ($\frac{1}{3}$ or 1:3), with a span of 3 m will have a rise of 1 m.

Pitch is dependent on the type and size of the materials used for the weatherproof covering and the building's exposure to windy conditions. They may be classified as either:

- Flat roofs where the pitch of the roof surface does not exceed 10°.
- Pitched roofs where the pitch of the roof surface exceeds 10°.

▲ **Figure 2.1** Roof pitch

Flat roofs are normally weathered using either rolls of bituminous felt built up in layers and bonded to the decking with cold or hot applied bitumen; or mastic asphalt hot applied in layers on a sheathing felt that covers the decking. In order to reflect some of the sun's heat both types of weathering may be protected with a heat reflecting paint or covered with a layer of pale coloured stone chippings.

Pitched roofs can be covered using clay or concrete tiles; quarried or man-made slates; wood shingles (normally cedar); thatch (reeds); and various profiled sheet materials manufactured from protected metals, plastics and cement/mineral fibre mix. In general as the pitch of the roof lowers the unit size of the covering material must increase.

Roof shape

Roofs may be further divided into a number of basic forms by their shape, the most common of which are illustrated in Figure 2.2.

Lean-to and mono-pitch roofs. Both have a single sloping surface. The lean-to roof abuts a higher wall or building and is often used for extensions, whereas mono-pitched roofs are normally free-standing.

Gable-end roofs have two sloping surfaces (double pitched) terminating at one or both ends with a triangular section of brickwork.

Hipped-end roofs are again double-pitched, but in these cases the roof slope is returned around one or both the shorter sides of the building to form a sloping triangular end.

Mansard roofs are double-pitched roofs, where each slope of the roof has two pitches. The lower part has a steep pitch (to act as walls) and often incorporates dormer windows for room in the roof applications whilst the upper part (acting as the main weathering) rarely exceeds 30°. The shorter sides of the building may be finished with either gable or hipped ends.

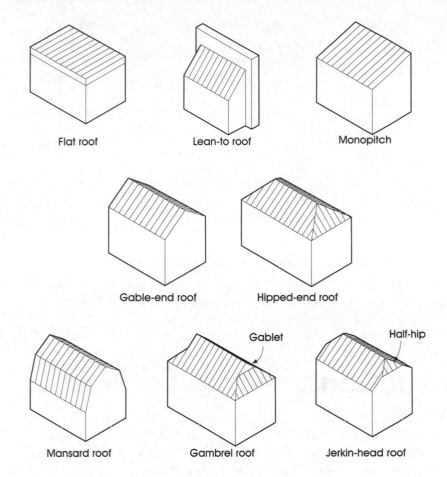

Flat roof   Lean-to roof   Monopitch

Gable-end roof   Hipped-end roof

Gablet   Half-hip

Mansard roof   Gambrel roof   Jerkin-head roof

◀ **Figure 2.2** Roof shapes

Gambrel roofs are double-pitched roofs incorporating a small gable or gablet at the ridge and a half-hip below. The gablet may be finished with tile hanging, timber cladding or a louvred ventilator.

Jerkin-head roofs are double-pitched roofs, which are hipped from the ridge part way to the eaves, and the remainder gabled. These are also termed as Dutch hipped roofs.

Roof elements

Refer to Figure 2.3:

- *Abutment.* The intersection where a lean-to or flat roof meets the main wall structure.
- *Eaves.* The lower edge of a roof, which overhangs the walls and where rainwater is discharged normally into a gutter. May be finished with a fascia board and soffit.
- *Gable.* The triangular upper section of a wall that closes the end of a building with a pitched roof.
- *Hip.* The line between the ridge and eaves of a pitched roof, where the two sloping surfaces meet at an external angle.
- *Parapet.* The section of a wall that projects above and is terminated some way beyond the roof surface.
- *Ridge.* The horizontal line of intersection between two sloping roof surfaces at their highest point or apex.

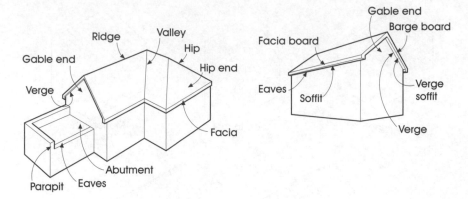

▶**Figure 2.3** Roof elements

- *Valley.* The line between the ridge and eaves of a pitched roof, where the two sloping surfaces meet at an internal angle.
- *Verge.* The termination or edge of a pitched roof at the gable or the sloping edge (non-drained) of a flat roof. May be finished with a bargeboard and soffit.

Flat roof construction

Timber flat roofs are similar in construction to that of upper floors. Their means of support, tying in/down, stiffening with the use of strutting and the decking all follow on similar lines.

Although the surface of the roof is flat, it is not horizontal. It should have a slope or fall on its top surface to ensure that rainwater will be quickly cleared and not accumulate on the roof. There are three ways of forming the slope as illustrated in Figure 2.4:

1. *Level joists:* The joists are laid level with the slope being formed by long tapering wedged shaped pieces of timber (firring pieces) being nailed to their upper edges. This is the most common method as it forms a level undersurface for fixing the ceiling plasterboard.

2. *Sloping joists:* The joists themselves are laid to the required slope and therefore no firrings are required. However it has the disadvantage of forming a sloping ceiling soffit.

3. *Diminishing firrings:* The joists are laid level at right angles to the slope and reducing or diminishing sections of timber are nailed to their top edge to create the fall. This method is rarely seen, but it is a useful method when timber boards are used for the roof decking, as these should be fixed parallel to the roof slope.

▲ **Figure 2.4** Methods of forming flat roof slopes

Flat roof carcass

Joist size and spacing. The size of flat roof joists can be found in the same way as floor joists (Figure 2.5). For new work the sectional size and spacing will always be specified. The Building Regulations contain tables for use in different situations, or alternatively they may be determined by calculation. Slightly longer spans are permitted for flat roof joists than floor joists of the same sectional size, as flat roofs

Short joists
forming verge

Tapered firring
to form slope

Joist

Ends built in main
wall at abutment

Strut

Ends supported on wall plate

Ends of joists overhang
to form eaves

Wall plate

◀ **Figure 2.5** Flat roof carcass

are only normally walked upon for maintenance purposes. The actual spacing of the joists must be related to the type of material used for the decking. Table 2.1 can be used as a guide.

Layout of joists. The positioning and levelling of flat roof joists is similar to the positioning and levelling of floor joists, except that flat roofs only occasionally have to be trimmed around openings, e.g. roof lights, chimney stacks, etc. Flat roofs, like floors, also require some form of strutting at their mid-span or at 1.8-m centres.

Joist supports. The joists are supported at the eaves and or verge depending on the joist's direction of span, either directly on the inner leaf of the cavity wall or a timber wall that has been secured on top of the inner leaf, see Figure 2.6. Where the ends of the joists abut a wall either at the main building or at a parapet various methods can be used for support including building in, joist hangers, timber wall pieces and steel angle sections.

Anchoring joists. In order to prevent strong winds lifting the roof, the joists must be anchored to the walls at a maximum of 2-m centres. This anchoring must be carried out to the joists that run both parallel with or at right angles to the wall. Figure 2.7 shows how a flat roof joist can be anchored to the wall using an anchoring or tie-down strap. The strap is nailed to the side of the joist and fixed to the wall by either nailing or plugging and screwing.

Single or double flat roofs. Binders may be incorporated in large flat roofs, either to provide intermediate support to long joists or to enable the roof to slope in both directions as illustrated in Figure 2.8. The use of a binder creates what is termed as a double flat roof. This reduces the effective span of the joists and enables the use of smaller

Table 2.1 Maximum clear span of flat roof joists and decking (m)

Based on the use of C16 grade softwood with dead load of up to $0.5\ kN/m^2$ with access only for maintenance and repair purposes.

Size of joist (mm) B × D	Joist spacing (centre/centre) 400 mm	450 mm	600 mm
38 × 97	1.74	1.72	1.67
38 × 122	2.37	2.34	2.25
38 × 147	3.02	2.97	2.85
38 × 170	3.63	3.57	3.37
38 × 195	4.30	4.23	3.86
38 × 220	4.94	4.76	4.34
50 × 97	1.97	1.95	1.89
50 × 122	2.67	2.64	2.53
50 × 147	3.39	3.34	3.19
50 × 170	4.06	3.99	3.69
50 × 195	4.79	4.62	4.22
50 × 220	5.38	5.19	4.74
75 × 122	3.17	3.12	3.00
75 × 147	3.98	3.92	3.64
75 × 170	4.74	4.58	4.19
75 × 195	5.42	5.23	4.79
75 × 220	6.07	5.87	5.38

Decking material	Finished thickness	Span (joist spacing)
Softwood T & G	16 mm	up to 500 mm
Softwood T & G	19 mm	up to 600 mm
Roofing grade chipboard	18 mm	up to 500 mm
Roofing grade chipboard	22 mm	up to 600 mm
Orientated strand board OSB	15 mm	up to 500 mm
Orientated Strand Board OSB	18 mm	up to 600 mm
External Plywood WBP	16 mm	up to 500 mm
External Plywood WBP	19 mm	up to 600 mm
Wood Wool Slabs	50 mm	up to 600 mm

Note: Recommended span of decking may vary between manufacturers of panel products.

Joist supported on inner leaf of wall at eaves or verge

Joist

Joist supported on wall plate at eaves or verge

Joist

Wall plate

Built in at abutment

Joist

Joist hanger at abutment

Joist

Timber wall piece at abutment

Joist

Bolted to wall

Steel angle at abutment

Joist

Bolted to wall

▲ **Figure 2.6** Joist supports

Joist

Wall plate

Framing anchor or truss clip

Tie down strap over wall plate and screwed to wall

Joist

Twisted tie down strap screw-fixed to joist and wall

◀ **Figure 2.7** Anchoring joists

Slope

Single roof

Slope   Slope

Double roof

Firrings

Joist

Joist

Saddle (double) joist hanger

Binder

◀ **Figure 2.8** Single and double flat roofs

sectional sizes. In addition the required width of verge and eave finishes are reduced.

Flat roof decking. Boarding or decking to flat roofs should be carried out just before they are to be waterproofed. Where timber boards are used they should be laid either with, or at a diagonal to the fall of the roof's surface. This is because any cupping or distortion of the boards can lead to pools of water being trapped in hollows formed on the roof's surface if boards are laid at right angles to the fall. See Figure 2.9. Chipboard roof sheets are available pre-covered with a layer of felt to provide some measure of initial protection.

▲ **Figure 2.9** Timber boarding for flat roofs

Metal-covered flat roofs. When sheet metal such as lead, copper or zinc is used as a flat roof covering material, wood-cored rolls in the direction of the fall and stepped drips across the fall must be incorporated in the roof construction. This is to enable sheets to be joined and also limit the size of the sheets to combat the effects of expansion in warm weather. Figure 2.10 shows a part plan and section through a metal-covered flat roof with rolls, drips and a box gutter.

▶**Figure 2.10** Section through metal-covered flat roof

Verge and eaves finishes

The finishing of verge and eaves to flat roofs is a similar process to that of pitched roofs. Both can be finished as either flush or overhanging with details as shown in Figure 2.11.

▲ **Figure 2.11** Flush and overhanging eaves

A drip batten is fixed at the lower eaves to the top of the fascia to extend the roof edge into the gutter (see Figure 2.12). This extension enables the roofing felt to be turned around it. Rainwater flows off the drip batten into the centre of the gutter, thus ensuring efficient discharge and reducing the risk of rot damage to timber. A drip batten is also fixed at higher eaves and verge edges, again to enable roofing felt to be turned around it. Any moisture is then allowed to drip clear of the fascia and not creep back under the felt by capillary attraction, with the subsequent likelihood of rot.

▲ **Figure 2.12** Use of drip batten to eaves of flat roof

Due to the depth of joists deep fascia boards are often required. These may be formed from solid timber, plywood (WBP weather and boil proof) or matchboarding. An alternative is to reduce the depth of the joists at the ends by either a splay or square cut as shown in Figure 2.13.

▲ **Figure 2.13** Reducing the depth of the joist

▲ Figure 2.14 Boxed gutter to a flat roof

Figure 2.14 shows how a boxed or internal gutter can be formed at the eaves as an alternative to external gutters. The gutter fall (1:60) can be achieved by progressively increasing the depth of cut-out in the joist ends towards the outlet.

Angle fillets, as shown in Figure 2.15, are fixed around the upper eaves, verge and edges of the roof which abut brickwork. They enable the roofing felt to be gently turned at the junction; sharp turns on felt lead to cracking, subsequent leaking and risk of rot. In addition the use of angle fillets also prevents rainwater from dripping or being blown over the edges of the roof.

▲ Figure 2.15 Use of angle fillets

Thermal insulation. There are two methods by which a flat roof can be insulated as shown in Figure 2.16.

■ *Insulation in the roof space* known as cold roof construction, this has thermal insulation and a vapour check at ceiling level. The roof space itself is cold and must be vented to the outside air to prevent interstitial condensation.

■ *Insulation above the roof space* known as warm roof construction, this has its thermal insulation and vapour barrier placed over the roof decking. The roof space is kept warmer than the outside air temperature and does not require ventilation.

The actual thickness of insulation used will be dependent on its type, positioning and the use of the building. The Building Regulations should be consulted for further information.

Ventilation. To reduce the likelihood of condensation within the roof space, ventilation is required by the Building Regulations. All cold flat roofs must be cross-ventilated at verge or eaves level by permanent vents. These must have an equivalent area equal to a continuous gap along both sides of the roof of 25 mm. The void in the roof between the insulation and the underside of the decking should be a minimum of 50 mm to allow free air movement. See Figure 2.17. Joists that run at right angles to the airflow should be counter-battened before decking to provide the free air space.

Cold deck flat roof construction

Labels: Returned eaves or verge; Ventilation above insulation; Wall and ceiling insulations to join up; Slope; Roofing felt with reflective chippings; Decking; Plaster board ceiling with vapour check; Insulation between joists; Joist and firring

Warm deck flat roof construction

Labels: Eaves; Slope; Roofing felt with reflective chippings; Insulation over decking; Vapour barrier; Joist and firring; Plaster board ceiling; Wall insulation carried up to underside of decking

◀ **Figure 2.16** Flat roof thermal insulation

This ventilation requirement can be achieved by either by:

■ Leaving a gap between the wall and soffit (this may be covered with a wire mesh to prevent access by birds, rodents and insects, etc.)

■ The use of proprietary ventilation systems in the soffit.

Where cross-ventilation is not possible due to abutting walls a warm roof construction should be used.

Ventilation opening equal to a continious gap of 25 mm

Counter batten

Cross – ventilation

At least 50 mm clear airway

Joist

25 mm

Joist at right– angles to air flow

Counter battens used to provide free air space

25 mm

Depth of joist

Cross – ventilation

At least 50 mm clear airway

25 mm

Joist parallel to air flow

▲ **Figure 2.17** Flat roof ventilation

Pitched roof construction

Timber pitched roofs may be divided into two broad but distinct categories:

- *Traditional framed cut roofs.* Almost entirely constructed on-site from loose timber sections and utilising simple jointing methods.

- *Prefabricated trussed rafters.* Normally manufactured under factory conditions.

Roof component terminology

The main structural component parts of a roof are illustrated in Figure 2.18.

- *Common rafters.* the main load bearing timbers in a roof, which are cut to fit the ridge and birdsmouthed over the wall plate,

- *Ridge.* The backbone of the roof, which provides a fixing point for the tops of the rafters, keeping them in line.

- *Wall plate.* Transfers the loads imposed on the roof, uniformly over the supporting brickwork. It also provides a bearing and fixing point for the feet of the rafters.

- *Hip rafter.* Used where two sloping roof surfaces meet at an external angle. It provides a fixing point for the jack rafters and transfers their loads to the wall.

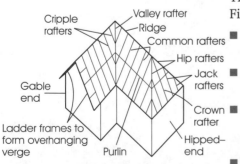

Cripple rafters
Valley rafter
Ridge
Common rafters
Hip rafters
Jack rafters
Gable end
Crown rafter
Ladder frames to form overhanging verge
Purlin
Hipped– end

▲ **Figure 2.18** Traditional roof terminology

- *Jack rafters.* Span from the wall plate to the hip rafter, like common rafters that have had their tops shortened.

- *Crown rafter.* A common rafter that is used in the centre of a hip end.

- *Cripple rafters.* Span from the ridge to the valley, like common rafters that have had their feet shortened (the reverse of jack rafters).

- *Valley rafter.* Like a hip rafter, but used at an internal angle. It provides a fixing point for the cripple rafters and transfers their loads to the wall.

- *Purlin.* A beam that provides support for the rafters in their mid span. Like a binder used in upper floor construction.

- *Ladder frame.* Also known as a *gable ladder* and is fixed to the last common rafter to form the overhanging verge on a gable roof. It consists of two rafters with noggins nailed between them.

- *Ceiling joists.* On which the ceiling plasterboard is fixed. They also act as ties for each pair of rafters at wall plate level.

- *Binders and hangers.* Stiffen and support the ceiling joists in their mid span to prevent them from sagging and distorting the ceiling.

- *Struts.* Transfer the loads imposed by the purlins onto a load-bearing partition wall.

- *Spreader plate.* Provides a suitable bearing for the struts at ceiling level.

- *Collar tie.* Used to prevent the spread of rafters in the same way as ceiling joists and may also be used to provide some support for the purlins.

Traditional roof structures

Traditional framed cut roofs may be constructed as single, double, or triple roofs, according to their span.

Single roofs

The rafters of all forms of single roof do not have any intermediate support (see Figure 2.19). Single roofs are only suitable for spans up to 5.5 m, as it is not economically viable to use increased sectional sizes of timber required for larger spans.

- *Couple roofs* consist of pairs of rafters fixed at one end to the wall plates and at the other to the ridge board. The span is restricted as, without a tie to the feet, the forces, which act on the roof, have a tendency to spread the walls.

- *Close couple roofs* are similar to the couple roofs, but with the feet of the rafters closed with a tie. This enables the span to be increased. The tie also acts as the ceiling joists. In addition a central binder may be fixed to the ceiling joist, hung from the ridge at every third or forth rafter spacing, to bind together and prevent any sagging in the ceiling joists.

- *Collar tie roofs* are similar to the close couple roofs, but in order to increase the span the tie is moved up the rafter a maximum of one third of the rise.

▲ **Figure 2.19** Single roof structures

Double roofs

The rafters of double roofs are of such a length that they require an intermediate support by purlins in their mid span, see Figure 2.20. This effectively reduces the span of the rafters and limits their size to economical sections. Struts, collars and hangers are spaced evenly along the roof at every third or fourth pair of rafters, to provide intermediate support for the purlins. Binders may also be fixed to the ceiling joist, hung from above to bind together and prevent any sagging in the ceiling joists.

▲ **Figure 2.20** Double roofs

Triple roofs

These use fabricated roof trusses spaced at intervals along a roof to provide intermediate support for the purlins, which in turn provide intermediate support for the common rafters. In common with triple floors they have been largely obsolete for many years, but you may still come across them in maintenance or renovation work (see Figure 2.21). Early trusses were mortised and tenoned together from fairly large section timbers. Later bolted trusses were developed; these were pre-fabricated in a workshop using simple overlapping joints, which are secured with bolts and timber connectors.

▲ **Figure 2.21** Triple roof

Pitched roof ends and intersections

Gables and hips

The ends of pitched roofs may be terminated with a gable wall, a sloping hip end roof or a combination of both as in gambrel and jerkin roofs. Figures 2.22 and 2.23 show typical details of gable and hip ends.

A **saddle board** is sometimes fixed to the common rafters as a bearing for the crown and hip rafters see Figure 2.24. The order of fixing can vary: either the crown is fixed first and the hips fitted in between; or the hips are fitted first and the crown is fitted in between.

An **angle tie** (Figure 2.25) traditionally termed as a **dragon tie** may be fitted across the wall plate to strengthen the joint and provide an additional bearing to take the thrust of the hip.

▶**Figure 2.22** Hip-end roof detail

▲ **Figure 2.23** Gable-end roof detail

▲ **Figure 2.24** Use of a saddle board at hip/ridge intersection

Valleys

Where two pitched roofs intersect at an internal angle a valley is formed between the two sloping surfaces. This valley may be constructed in one of two ways. Either a valley rafter is used and the feet of the common rafters (cripple rafters) of both roofs are trimmed into it, or the rafters

Cogged over tie
Birds mouthed over wall plate
Hip rafter
Angle tie
Wall plate
Corner of wall plate removed to provide bearing for hip

Hip rafter
Angle tie
Wall plate

◀ **Figure 2.25** Use of angle tie at hip across wall plates

of one roof are run through and lay boards are used to take the feet of the cripple rafters of the other roof. See Figure 2.26.

Cripple rafters
Valley rafter

Layboard
Cripple rafters

▲ **Figure 2.26** Valley designs

Parapet gutter is where a pitched roof terminates at the eaves by a parapet wall. Support and framework for a lead- or zinc-lined parapet gutter must be formed. See Figure 2.27.

Trimming to openings

Where openings occur in roofs, in either the rafters or ceiling joists or both, these have to be trimmed. Framing anchors or housing joints are used to join the trimmers, trimmings and trimmed components together. Figure 2.28 gives a section through a ceiling hatch showing the trimmed

tilting fillet
lay board
tilting fillet
rafter
gutter board
brackets fixed to side of each rafter

▲ **Figure 2.27** Parapet gutter

Hatch lining
Plywood hatch cover
Housing, joist hanger or framing anchors
Plasterboard Architrave Planted stop Joist
Section

◀ **Figure 2.28** Trimming to loft hatch

opening and hatch lining. Figure 2.29 gives a part plan and section through a roof showing the trimming around a chimney stack and the back gutter detail.

▲ **Figure 2.29** Trimming to chimney stack

To comply with the Building Regulations, no combustible material, including timber, is to be placed within 200 mm of the inside of the flue lining; or, where the thickness of the chimney surrounding the flue is less than 200 mm, no combustible material must be placed within 40 mm of the chimney. See Figure 2.30.

▶**Figure 2.30** Separation of flue and combustible materials

Water tank platforms

These are ideally situated centrally over a load-bearing wall. The platform should consist of boarding supported by joists laid on top of the ceiling joists and at right angles to them. Figure 2.31 shows a typical water tank platform.

Anchoring roofs

Rafters can be skew nailed to the wall plates or, in areas noted for high winds, framing anchors or truss clips where appropriate can be used. See Figure 2.32. Wall plates must be secured to the wall with straps at 2 m centres. The rafters adjacent to the gable end should also be tied into the wall with straps at 2-m centres. The rafters and ceiling joists adjacent to the gable end should also be tied into the wall with metal restraint straps at 2-m centres.

▲ **Figure 2.31** Water tank platform

◀ **Figure 2.32** Anchoring of rafters and wall plates

Roof sizes and spacing

Size and spacing of members are dependent on the location, pitch and span of the roof. For new work, the sectional size and spacing to use will always be specified. In other situations, Table 2.2 can be used as a guide. Further tables are contained in the Building Regulations AD:A.

Determining lengths and bevels

There are three main methods that can be used to determine the lengths and bevels required for a traditional cut roof:

- Full size setting out, only really suitable for fairly small rise/span gable end roofs.
- Scale drawings and geometrical development. This chapter covers the use of this information; the actual geometry is covered in Book 1.
- Roofing square.

Table 2.2 Maximum clear span of pitched roof rafters and ceiling joists (m)

Size of rafter (mm) B × D	Joist spacing (centre to centre) 400 mm	450 mm	600 mm
38 × 100	2.28	2.23	2.10
38 × 125	3.07	2.95	2.69
38 × 150	3.67	3.53	3.22
50 × 100	2.69	2.59	2.36
50 × 125	3.35	3.23	2.94
50 × 150	4.00	3.86	3.52

Size of ceiling joist (mm)			
38 × 72	1.11	1.10	1.06
38 × 97	1.67	1.64	1.58
38 × 122	2.25	2.21	2.11
38 × 147	2.85	2.80	2.66
50 × 72	1.27	1.25	1.21
50 × 97	1.89	1.86	1.78
50 × 122	2.53	2.49	2.37
50 × 147	3.19	3.13	2.97

Based on using C16 grade softwood rafters for roofs over 30° and up to 45° pitch, with a dead load of up to 0.5 kN/m², imposed loading of 0.75 kN/m² suitable for most sites located up to 100 m above ordnance datum with access only for maintenance and repair. Ceiling joists are for dead loads between 0.25 and 0.50 kN/m².

Full size setting out

This method involves setting out in full size on a suitable surface, such as a concrete base, a couple of sheets of plywood, or couple of scaffold boards or joists positioned at right angles to each other.

For practical purposes the initial setting out is done using single lines, without the width or thickness of the members being shown, see Figure 2.33. Lines to indicate the rise and run (half the roof span) of the rafter are drawn first, followed by the diagonal rafter line (pitch

◀ **Figure 2.33** Marking pattern rafter from full size setting

line). This will indicate the required plumb and seat cuts as well as the theoretical or true length of the rafter. The end sections of the wall plate and the ridge can be added and then the width of the rafter, two thirds above and one third below the pitch line.

A pattern rafter can then be laid on top of this and marked out. The rest of the rafters can then be marked out and cut from the pattern.

Scale drawings and geometrical development

After having produced your own scale drawings and geometrical developments, or a setter-out has produced them, either manually or with the aid of a computer drawing programme (CAD), you will have to interpret the information in order to mark out the roof members.

Figure 2.34 shows a typical scale drawing of a part plan and section of a hipped end roof with purlins. The drawing shows all the developments, angles and true lengths required to set out and construct the roof. The same plumb and seat cuts indicated for the common rafter are used for the crown, jack and cripple rafters. However those of the hip rafter will be different due to its increased run.

An adjustable bevel square can be set up to the required angles by placing over the actual drawing as shown in Figure 2.35.

The true lengths shown on the drawing are measured on the pitch line which is a line marked up from the underside of the common rafter, one third of its depth. As the hip and valley rafters are usually of deeper section, the pitch line on these is marked down from the top edge at a distance equal to two thirds the depth of the common rafter as shown in Figure 2.36.

Take care when scaling the dimensions from the drawing as 1 mm difference on a 1:20 drawing is 20 mm difference at full size. It's best always to cut a pair of rafters and try them up on the roof for size, before progressing onto cutting the full requirement. Figure 2.37 illustrates the effect of any possible errors in the length of rafters.

The true length of the common and hip rafters is measured on the pitch line from the centre line of the ridge to the outside edge of the wall plate. For jack rafters, it is from the centre line of the hip to the outside

▲ **Figure 2.34** Roofing angles and true lengths

edge of the wall plate; for the cripple rafter it is from the centre line of the ridge to the centre line of the valley.

Therefore when marking out the true lengths of the roofing components from the single line drawing, an allowance in measurement must be made. This allowance should be an addition for the eaves overhang and a reduction to allow for the thickness of the components. If this reduction is not apparent, it may be found by drawing full size the relevant intersecting components.

◀ **Figure 2.35** Setting an adjustable bevel for a rafter plumb cut

▲ **Figure 2.36** Setting out rafters

◀ **Figure 2.37** Possible effect of scaling errors

Figure 2.38 shows the intersection between the ridge, common, crown and hip rafters. The reduction of the hip rafters is shown. The reduction for the common rafters is always half the ridge thickness and for the crown rafter half the common rafter thickness. These reductions should be marked out at right angles to the plumb cut. When a saddle board is fitted, its thickness must be allowed for when making the reductions to the hips and crown.

Figure 2.39 shows details at the corner intersection of the wall plates and the hip rafter, as the corner of the wall plates has been cut off to provide a flat surface for the hip plumb cut, the plumb cut itself must be moved forward to fit.

▲ **Figure 2.38** Reductions for material thickness

Corner of wallplate
removed to thickness
of hip

Hip plumb
cut

Pitch line

Birdsmouth

Half hip
removed

Half hip set forward

▶ **Figure 2.39** Hip rafter
birdsmouth

Roofing square

Setting out rafters

After having studied geometrically in Book 1 the lengths and angles
required for hipped, double and valley roofs, the use of the steel square
to find the same lengths and angles should be fairly straightforward, as
it is merely the application of the geometric principles.

Most roofing squares contain sets of tables on them, which give rafter
lengths per metre run for standard pitches, although in practice these
tables are rarely used.

The wide part of a roofing square is the blade and the narrow part, the
tongue (see Figure 2.40). Both the blade and the tongue are marked out
in millimetres. Most carpenters will make a fence for themselves using
two battens and four small bolts and wing nuts.

Fence

Tongue

Blade

▲ **Figure 2.40** Roofing square and
fence

To set out a roof using the roofing square, the rise of the roof is set on
the tongue and the run of the rafter is set on the blade (run of rafter =
half of the span). In order to set the rise and run on the roofing square,
these measurements must be scaled down and it is usual to divide them
by ten.

> ### *Example*
>
> For a roof with a rise of 2.5 m and a rafter run of 3.5 m, the scale lengths to
> set on the roofing square would be:
> Rise 2.5 m ÷ 10 = 250 mm; run 3.5 m ÷ 10 = 350 mm

Figure 2.41 shows how to set up the roofing square and fence to obtain
the required lengths and angles. The lengths will, however, be scale
lengths.

Figure 2.42 shows how the roofing square may be stepped down the
rafter ten times to obtain its actual length. Alternatively, the scale
length can be measured off the roofing square and multiplied by ten to
give its actual length.

Allowances. The roofing square, like the geometrical method, gives
the true lengths of members on the pitch line. Therefore the same

▲ **Figure 2.41** Setting up a roofing square to obtain various scale lengths and angles

▲ **Figure 2.42** Using a roofing square

allowances in measurement as stated before must be made. The length for the shortest jack rafter can be found by dividing the length of the common rafter by one more than the number of jack rafters on each side of the hip. This measurement is then added to each successive jack rafter to obtain its length.

Example

For the roof shown in Figure 2.43 with three jack rafters on each side of the hip and a common rafter length of say 2.1 m:
Length of short jack (1) = 2.1 m ÷ 4 = 525 mm
Length of middle jack (2) = 525 mm + 525 mm = 1.05 m
Length of long jack (3) = 1.05 m + 525 mm = 1.575 m

The true lengths and angles for valley and cripple rafters can be found by using the same methods as used for the hip and jack rafters.

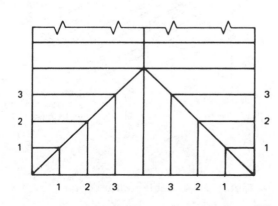

▶**Figure 2.43** Calculating jack rafter lengths

Roof erection

The procedure used for erection is similar for most types of traditional cut roof.

Gable roof

The procedure of erection for a gable-end single roof would be as follows (see Figure 2.44):

1. The wall plate, having been bedded and levelled by the bricklayer, must be tied down.

2. Mark out the position of the rafters on the wall plate and transfer these onto the ridge.

3. Place ceiling joists in position and fix by nailing to the wall plate; a temporary working platform inside the roof space can be created by placing scaffold boards over the ceiling joists.

4. Make up two temporary A-frames. These each consist of two common rafters with a temporary tie joining them at the top, leaving a space for the ridge and a temporary tie nailed to them in the position of the ceiling joists for ease of handling.

5. Stand up the A-frames at either end of the roof, fix in position by skew nailing to the wall plate and the ceiling joists.

6. Fix temporary braces to hold the A-frames upright.

7. Insert the marked out ridge board and nail in position through the top edge of the A-frame rafters.

8. Progressively position pairs of rafters along the roof, securing them by nailing them at both the ridge and wall plate. Diagonal rafter bracing may be required to stabilise the roof in windy locations.

9. Add noggins to form a gable ladder where an overhanging verge is required.

▲ **Figure 2.44** Gable roof erection

10. Allow bricklayers to build the gable end walls.

11. Finish the roof at the verge and eaves with bargeboard, fascia and soffit as required. (This stage is covered later in this chapter).

The erection of double-cut roofs follows the same procedure, except for the positioning and fixing of the purlin and any associated struts and binders. These may be fitted either after the A-frames and a central pair of rafters have been fixed, or left until all the common rafters have been fixed. The former method is often preferred, especially for longer spans as an early fixed purlin provides a point on which to slide the intermediate common rafters up to the ridge and eases any potential problems associated with fixing to a run of possibly sagging rafters.

Hipped-end roof

The procedure of erection for a hipped-end double roof would be as follows (see Figure 2.45):

1. The wall plate, having been bedded and levelled by the bricklayer, must be tied down.

2. Mark out the position of the rafters on the wall plate and transfer these onto the ridge.

3. Place the ceiling joists in position and fix by nailing to the external and any internal wall plates, a temporary working platform inside

▲ **Figure 2.45** Hip-end roof erection

the roof space can be created by placing scaffold boards over the ceiling joist. Where additional height is required to reach the ridge, a temporary scaffold from the floor below may need erecting.

4. Make up two temporary A-frames. These each consist of two common rafters with a temporary tie joining them at the top, leaving a space for the ridge and a temporary tie nailed to them in the position of the ceiling joists.

5. Stand up the A-frames towards either end of the roof at the position of the last common rafter (half span of roof from corner of wall plate to centre line of rafter), fix in position by skew nailing to the wall plate and the ceiling joists.

6. Fix temporary braces to hold the A-frames upright.

7. Insert the marked-out ridge board and nail in position through the top edge of the A-frame rafters.

8. Fix the crown and hip rafters.

9. Fix the purlins, struts and binders.

10. Fix the remaining common rafters and jack rafters. Short return ceiling joists are fixed to the feet of the crown and jack rafters at the hipped ends.

11. Fix the collars and the hangers.

12. Finish the roof at the eaves with fascia and soffit as required.

Prefabricated trussed rafters

Prefabricated roof components developed because of the need to:

■ Reduce the amount of timber in a roof.

■ Simplify the often-complicated on-site cutting of a roof.

■ Enable large spans without the need for intermediate supporting walls

The earlier designs were bolted roof trusses, which developed from the traditional jointed king and queen post trusses. Then came nail plate trussed rafters, which are almost exclusively used today for new work. The essential difference between the two is that in a trussed rafter roof, every rafter is trussed and carries only its own proportion of the total load, whereas the older bolted roof trusses supported a purlin, which in turn carried the roof loads from a number of adjacent common rafters.

Nail plate trussed rafters

These are prefabricated by a number of specialist manufacturers in a wide range of shapes and sizes. Figure 2.46 shows a range of standard trussed rafter configurations. They consist of prepared timber laid out in one plane, with their butt joints fastened with nail plates. Figure 2.47 shows a typical design for a trussed rafter.

In use the trusses are spaced along the roof at between 400 mm and 600 mm centres and fixed to the wall plate, preferably using truss clips or tie down straps, see Figure 2.48. Unlike common rafters the

'W' (fink)

Mono

Fan

Scissors

King post

Dormer

Attic

Howe

Asymmetric

Truss or
laminated purlin

Cross wall

Min. load to eaves

▲ **Figure 2.46** Standard trussed rafter configurations

▲ **Figure 2.47** Nail plate trussed rafter

Truss
clip

Tie down
strap

▲ **Figure 2.48** Fixing trussed rafters to wall plate

trusses simply rest on top of the wall plates and are not birdsmouthed over them.

In order to provide lateral stability the roof requires binders at both ceiling and apex level and diagonal rafter bracing fixed to the underside of the rafters (see Figure 2.49). These must be fixed in accordance with the individual manufacturer's requirements.

▲ **Figure 2.49** Bracing and binder details

Figure 2.50 shows how the gable wall must be tied back to the roof for support. This is done using lateral restraint straps at 2 m maximum centres both up the rafter slope and along the ceiling tie.

Roof end and intersections using trussed rafters

Prefabricated gable ladders as shown in Figure 2.51 can be fixed to the last truss when an overhanging verge is required. Bargeboards and soffits are in turn nailed directly to the ladder.

Where hip ends and valleys occur in trussed rafter roofs, these may be formed either by using loose timber and cutting normal hip and valley rafters, etc., in the traditional manner, or by using specially manufactured components.

Hip end. Figure 2.52 illustrates how a hip end may be formed using cut timber in the traditional way. These are pitched against a girder truss formed by nailing two or more trusses together.

Timber packing

maximum
2 m centres

Lateral restraint straps

Noggin

Strap fixed to
uncut block

Pack

Timber packs to
be a tight fit

Restraint strap
at rafter level

Pack

Noggin

Restraint strap
at ceiling level

▲ **Figure 2.50** Gable end details

▶**Figure 2.51** Prefabricated gable ladder

Gable ladder

Last truss

Standard
trussed
rafter

Traditional
cut rafter
and ceiling
joists

Ceiling joists
supported on girder
truss by hangers

Girder truss
(Multiple trussed
rafters joined
together)

◀ **Figure 2.52** Traditional cut rafter
hip end trussed rafter roof

65

Hip rafter

Compound hip girder trusses

Purlin

Jack rafter

Short return ceiling joist

Hip mono trusses

Ceiling joist

▲ **Figure 2.53** Hip end detail using mono trusses

Figure 2.53 illustrates how a hip end may be formed using a compound hip girder truss to support hip mono trusses. Loose hip rafters, infill jack rafters and ceiling joists must still be cut and fixed in the normal way.

Valley. Figure 2.54 shows how a valley may be formed where two roofs intersect at a tee junction, using diminishing jack rafter frames nailed on to lay boards. The ends of the rafters on the main roof are carried across the opening by suspending them from the compound girder truss using suitable joist hangers.

Water tank platforms. These should be placed centrally in the roof with the load spread over at least three trussed rafters. The lower bearers of the platform should be positioned so that the load is transferred as near as possible to the mid-third points of the span (see Figure 2.55).

Trimming. Wherever possible, openings in roofs for chimney stacks or loft hatches should be accommodated within the trussed rafter spacing. If larger openings are required the method shown in Figure 2.56 can be used. This entails positioning a trussed rafter on either side of the opening and infilling the space between with normal rafters, purlins and ceiling joists. For safety reasons, on no account should trussed rafters be trimmed or otherwise modified without the structural designers' approval.

Attic-trussed rafters

These are available to accommodate rooms in the roof space and are available in two types, each requiring a different approach, see Figure 2.57:

■ Trusses that include a large sectioned bottom tie that acts as the floor joist. The roof is normally erected in the same way as standard

Diminishing jack
rafter frames

Compound
girder trusses

Standard trusses
to main roof

Lay board

Girder trusses

Main trussed rafter

Support detail

▲ **Figure 2.54** Valley detail

Water tank

Built up platform

▲ **Figure 2.55** Water tank platform

▲ **Figure 2.56** Trimmings to openings

Large section tie acts as a floor joist

Floor joist

◄ **Figure 2.57** Attic-trussed rafters

trussed rafters, with the centre section then being boarded over to form the floor. However you may come across roofs that have been assembled and boarded out at ground level or even off-site in a factory and simply lifted in place.

■ The alternative is attic trusses that require an upper suspended floor to be formed at the eaves level first. The trusses are then erected and fixed to the floor joists, again with the centre section boarded over to form the floor.

Standard windows in the gable end wall can be used to provide much of the light and ventilation requirements, which can be supplemented by sloping roof lights positioned at intervals along the roof between the standard truss spacing.

Where **dormer windows** are required an opening larger than the truss spacing will be required. This can be achieved using a similar method to that of other trimmed openings, see Figure 2.58.

Pairs of attic trusses are nailed together and positioned at both ends of the required opening. The space in between is in-filled with either special trimmed trusses or normal cut rafters and joists. Again for safety reasons, on no account should trussed rafters be trimmed or otherwise modified without the structural designers' approval.

Elevation

- Beam to support ends of trimmed rafter
- Head of window opening
- Trimmed infill rafter
- Pair of attic trussed rafters nailed together
- Cill beam
- Centre section boarded out to form floor
- Floor joist

Section

- Framing for dormer roof
- Head of window opening
- Infill studding to side of dormer

Plan

- Beam
- Position of window and cill beam
- Pair of trussed double rafters
- Dormer window flat roof joists
- Trimmed rafters

▲ **Figure 2.58** Attic-trussed rafter roof dormer window details

Erection of trussed rafters

The main problem encountered with the erection of trussed rafters is handling. In order not to strain the joints of the trussed rafters they should be lifted from the eaves, keeping the rafter in a vertical plane with its apex uppermost. Inadequate labour or care will lead to truss damage. Trussed rafters may be lifted into position with the aid of a crane, either singly from the node joints using a spreader bar and slings, or in banded sets. In both cases these should be controlled from the ground using a guide rope to prevent swinging. In addition, where large cranes are available, the entire roof may be assembled, braced, felted and battened at ground level and then lifted into position as one whole weatherproof unit, requiring only tiling at a later stage. This method is particularly suited to the rapid erection of timber frame buildings. Figure 2.59 illustrates the handling of trussed rafters.

The erection procedure for a gable-end roof using trussed rafters is as follows (Figure 2.60):

1. Mark the position of the trusses along the wall plates.

Crane

Spreader bar

Slinging from joints

Guide rope

See-saw effect

Overstraining joints

Felted and battened roof assembled at ground level and lifted into place

▲ **Figure 2.59** Handling trussed rafters

Trussed rafter positions
marked on wall plate

Order of erection

Temporary brace

First trussed rafter

$\frac{1}{2}$ Span extent of
under-rafter bracing

Centre rafters erected
after end sections

$\frac{1}{2}$ Span extent of
under-rafter bracing

Temporary Ties

Repeat this erection
sequence on other
end of roof

▲ **Figure 2.60** Trussed rafter erection procedure

2. Once up on the roof, the first trussed rafter can be placed in position at the end of the under-rafter diagonal bracing. It can then be fixed at the eaves, plumbed and temporarily braced.

3. Fix the remaining trussed rafters in position one at a time up to the gable end, temporarily tying each rafter to the preceding one with a batten.

4. Fix diagonal bracing and binders.

5. Repeat the previous procedure at the other end of the roof.

6. Position and fix the trusses between the two braced ends one at a time, and fix binders.

7. Fix ladder frames and restraint straps.

8. Finish the roof at the eaves and verge with fascia, bargeboard and soffit as required.

Verge and eaves finishing

Verges

The termination or edge of a pitched roof at the gable end often overhang the wall and are finished with a bargeboard and soffit, see Figure 2.61.

▲ **Figure 2.61** Gable-end detail

To finish the overhanging verge of a gable end roof, the ridge and wall plate are extended past the gable end wall, and an additional rafter is pitched to give the required gable overhang. Noggins are fixed between the last two rafters to form a gable ladder. This provides a fixing for the bargeboard, soffit and tile battens.

Bargeboard

The continuation of a fascia board around the verge or sloping edge of the roof (typically from 25 mm × 150 mm planed-all-round PAR softwood) provides a finish to the verge. The lower end of the bargeboard is usually built up to box in the wall plate and eaves. A template may be cut for the bargeboard eaves shaping in order to speed up the marking out where a number of roofs are to be cut and also to ensure that each is the same shape (especially where more elaborate designs are concerned).

Two methods may be used for determining the bevels at the apex (top) and foot (bottom) of the bargeboard:

■ *Marking in position.* The board is temporarily fixed in position, so a spirit level can then be used to mark the plumb cut (vertical) and seat cut (horizontal) in the required positions (see Figure 2.62).

■ *Determining bevels.* Adjustable bevel squares may be set to the required angles for the plumb and seat cuts, using a protractor (Figure 2.63) or ones previously set from the scale geometrical drawings. These angles will be related to the pitch of the roof.

The sum of all three angles in a triangle will always be 180°. Therefore in a 30° pitched roof, the apex angle will be 120° (180° – 2 × pitch), making the plumb cut for each barge board 60° (half of apex angle). The seat cut is at right angles to the plumb cut and is at the same angle as the roof pitch, see Figure 2.64.

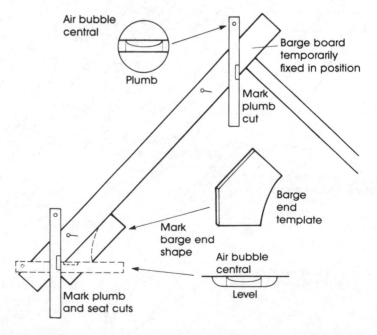

▲ **Figure 2.62** Marking out a bargeboard

▲ **Figure 2.63** Setting an adjustable bevel to a known angle

▲ **Figure 2.64** Determining angles for a bargeboard

Fixing the bargeboard

The foot of a bargeboard may be either mitred to the fascia board, butted and finished flush with the fascia board or butted and extended slightly in front of the fascia board (see Figure 2.65). The actual method used will depend on the specification and/or supervisor's instructions.

▲ **Figure 2.65** Jointing barge to fascia board

The mitred joint is preferred for high quality work. The angle of the mitre for the bargeboard and fascia is best marked in position. Temporarily fix each in position, one at a time. Use a piece of timber of the same thickness to mark two lines across the edge of the board and join the opposite corners to form the mitre as illustrated in Figure 2.66. The face angle will be 90° for the fascia board and a plumb cut for the bargeboard. Where timber of sufficient length is not available for a continuous barge, splayed heading joints may be used as shown in Figure 2.67.

After marking, cutting to shape, mitring and fitting, the bargeboard can be fixed to its gable ladder by double nailing at approximately 400 mm centres. Use either oval nails, wire nails, lost-head nails or cut nails. These should be at least $2\frac{1}{2}$ times the thickness of the bargeboard in length in order to provide a sufficiently strong fixing. For example, an 18-mm thick bargeboard would require nails of at least 45 mm long (50 mm being the nearest standard length). All nails should be punched below the surface, ready for subsequent filling by the painter.

▲ **Figure 2.66** Marking barge to fascia board mitre

▲ **Figure 2.67** Bargeboard lengthening/fixings

Eaves

The lowest part of a pitched roof slope where the ends of the rafters terminate usually overhang the wall and are finished with a fascia board and soffit.

Eaves may be finished as either (Figure 2.68):

- flush
- overhanging, open or closed
- sprocketed.

▲ **Figure 2.68** Eaves details

Flush eaves. In this method the ends of the rafters are cut off 10 to 15 mm past the face of the brickwork and the fascia board is nailed directly to them to provide a fixing point for the gutter. The small gap between the back of the fascia board and the brickwork allows for roof space ventilation.

Open eaves. These project well past the face of the wall to provide additional weather protection. The ends of the rafters should be prepared as they are exposed to view from the ground. In cheaper quality work the fascia boards are often omitted and the gutter brackets fixed directly to the side of the rafter.

Closed eaves. These overhang the face of the wall the same as open eaves except that the ends of the rafters are closed with a soffit. Cradling brackets are nailed to the sides of the rafters to support the soffit at the wall edge.

Sprocket piece. Flush, open and closed eaves often use sprocket pieces nailed to the top of each rafter to reduce the pitch of the roof at the eaves. This has the effect of easing the fast flowing rainwater under storm conditions into the gutter.

Sprocketed eaves. On steeply pitched roofs the flow of rainwater off the roof surface has a tendency to overshoot the gutter. Sprockets can be nailed to the side of each rafter to lower the pitch and slow down the rainwater before it reaches the eaves. This reduces the likelihood of rainwater overshooting the eaves and/or hitting the front of the gutter and splashing back soaking the eaves timbers, with the subsequent risk of rot. In addition the use of sprockets also enhances the appearance of a roof giving it a distinctive 'bell-cast' appearance.

Fascia board

The horizontal board (typically ex 25 mm × 150 mm PAR softwood) which is fixed to the ends of the rafters to provide a finish to the eaves and a fixing for the guttering. Before fixing the fascia board the rafter feet will require marking and cutting to plumb and line as shown in Figure 2.69. A seat cut may also be required depending on the assembly detail.

- Measure out from brickwork the required soffit width and mark on the last rafter at either end of the roof.
- Mark the plumb cut and the seat cut if required using a spirit level.
- Stretch a string line between the end two rafters and over the tops of the other rafters. Use a spirit level to mark each individual plumb cut.
- Cut the plumb cuts using either a handsaw or portable circular saw.
- Where a seat cut is required, move the line down to the seat cut position on the end rafters. Use a spirit level to mark each individual seat cut.
- Cut the seat cuts using either a hand or portable circular saw.

▲ **Figure 2.69** Marking out plumb cut at eaves

▲ **Figure 2.70** Jointing a fascia board over rafter end

Where timber of sufficient length is not available for a continuous fascia board, splayed heading joints may be used. Figure 2.70 shows how these should be positioned centrally over a rafter end.

Where level fascia boards are returned around corners it is standard practice to use a mitre at the external corners and butt at the internal corners. Both of these joints should be secured by nailing (50 mm ovals).

Fascia boards are fixed to the end of each rafter using two nails. These are normally either oval nails, wire nails, lost-head nails or cut nails, at least $2\frac{1}{2}$ times the thickness of the fascia board in length. Typically 50 mm or 62 mm nails provide a sufficiently secure fixing. All nails should be punched below the surface ready for subsequent filling by the painter.

Prior to final fixing, the fascia should be checked for line as shown in Figure 2.71:

■ Drive nails on the face of the fascia at each end of the roof. Strain a line between them.

■ Cut three identical pieces of packing, place one at each end under the line and use the third to check the distance between the fascia and the line at each rafter position.

■ Pack out or use a saw to ease ends of rafters as appropriate, so that the packing piece just fits between the fascia and the line.

▶**Figure 2.71** Checking fascia board for line

Soffit boards

These are used to close the gap between the fascia and the wall of the building; normally either strips of non-combustible sheet material, hardboard or formed from tongued and grooved matching (T & G).

Soffits normally tongue into the fascia board and are fixed to the underside of cleats or L-shaped brackets (cradling) which are themselves fixed to the sides of each rafter at the seat cut line or required soffit line.

Sheet material soffits are typically fixed using two 25-mm galvanised wire nails at each cleat or cradle position. Matchboarded soffits may be either surface nailed or secret nailed through the tongue using 38 mm oval nails at each cleat or cradle position.

Preservation, insulation and ventilation

Preservative treatment

It is recommended that all timber used for verge and eaves finishes is preservative treated before use. Any preservative-treated timber cut to size on site will require re-treatment on the freshly cut edges/ends. Applying two brush flood coats of preservative will achieve this.

Thermal insulation

This was normally done by placing insulation between the ceiling joists and incorporating a vapour check, such as foil-backed plasterboard, at ceiling level. However with the increased levels of thermal insulation now required for new buildings, a second layer of insulation is often laid over the ceiling joists (Figure 2.72).

Roof ventilation

Care must be taken not to block the eaves with the insulating material, as the roof space must be well ventilated in order to prevent any warm moist air that passes through the ceiling from condensing in the roof space.

To reduce the likelihood of condensation within the roof space ventilation is required by the Building Regulations. All roofs must be cross-ventilated at eaves level by permanent vents. These must have an equivalent area equal to a continuous gap along both sides of the roof of 10 mm or 25 mm where the pitch of the roof is less than 15° or the insulation follows the pitch of the roof. See Figure 2.73. This ventilation requirement can be achieved by:

- Leaving a gap between the wall and soffit (this may be covered with a wire mesh to prevent access by birds, rodents and insects, etc.).
- Using a proprietary ventilation strip fixed to the back of the fascia.
- Using proprietary circular soffit ventilators let into the soffit at about 400 mm centres.
- Lean-to roofs must have high-level ventilation in the roof slope close to the abutting wall. Proprietary vents are available for this purpose, these must have an equivalent area equal to a continuous gap along the abutment of 5 mm.
- Where the insulation follows the pitch of the roof, there must be at least a 50 mm air space between the insulation and the underside of the roof structure, and high-level ventilation at the ridge equal to a continuous gap of at least 5 mm. Proprietary ridge vents are available for this purpose.
- Roofs where the span exceeds 10 m or non-rectangular roofs may require additional ventilation, totalling 0.6% of the roof area.

▲ **Figure 2.72** Typical pitch roof insulation between and over ceiling joists

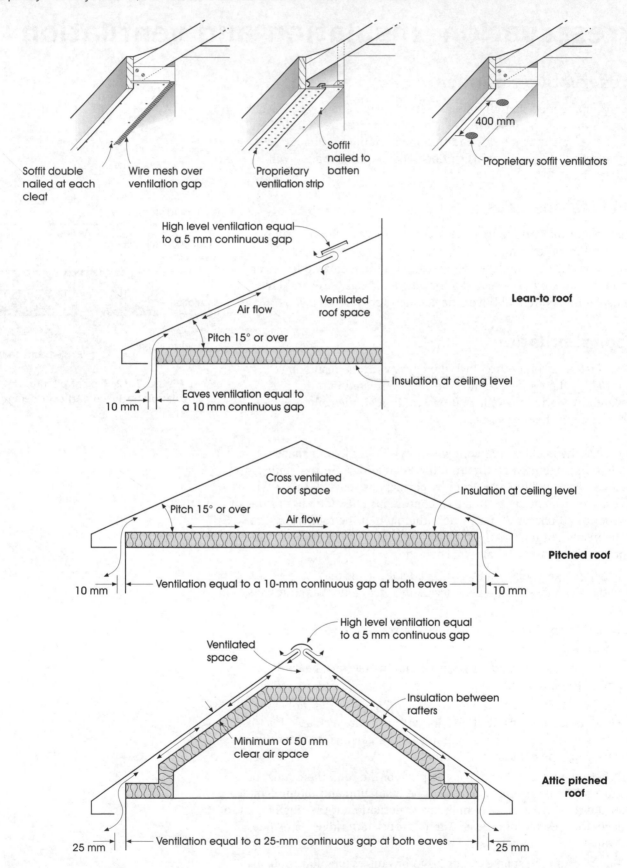

Soffit double nailed at each cleat

Wire mesh over ventilation gap

Proprietary ventilation strip

Soffit nailed to batten

400 mm

Proprietary soffit ventilators

High level ventilation equal to a 5 mm continuous gap

Ventilated roof space

Lean-to roof

Air flow

Pitch 15° or over

Insulation at ceiling level

10 mm

Eaves ventilation equal to a 10 mm continuous gap

Cross ventilated roof space

Insulation at ceiling level

Pitch 15° or over

Air flow

Pitched roof

10 mm

Ventilation equal to a 10-mm continuous gap at both eaves

10 mm

High level ventilation equal to a 5 mm continuous gap

Ventilated space

Insulation between rafters

Minimum of 50 mm clear air space

Attic pitched roof

25 mm

Ventilation equal to a 25-mm continuous gap at both eaves

25 mm

▲ **Figure 2.73** Pitched roof ventilation

Irregular plan roofs

In general most buildings are regular on plan and have equally pitched roofs, i.e. they have square corners and both sides are of equal slope. Sometimes, to create an architectural feature or to make use of a limited tapering site, buildings are constructed with an irregular plan. This therefore requires the use of irregular roof shapes and unequal pitches.

An equal-pitched hip-end roof on a building whose long sides are not parallel is illustrated in Figure 2.74. All of the rafters are kept square on plan to their respective wall plates, to avoid compound plumb and seat cuts to the common rafters. This necessitates a **split ridge** and a flat section to the top of the roof. The alternatives to this, which are impractical, would result in a sloping ridge and cut courses of roof tiles, or sloping eaves. Figure 2.75 shows a method of forming the split ridge, where the flat section is a maximum of about 300 mm. Two ridge boards can be used with bearers positioned between them at each rafter. For wider flat sections a tapering ridge framework should be used as illustrated in Figure 2.76.

▲ **Figure 2.74** Irregular plan roof

◄ **Figure 2.75** Split-ridge details

▲ **Figure 2.76** Split-ridge detail for wider flat section

▲ **Figure 2.77** Outline of traditional Mansard roof

Mansard roofs

These are constructed so that accommodation can be provided in the roof space. Traditionally the roof was constructed from framed Mansard roof trusses, which were a combination of a king post truss and a queen post truss, as illustrated in Figure 2.77. Today they will normally be formed using Mansard-trussed rafters.

A method of forming a Mansard roof using on-site cut timbers is illustrated in Figure 2.78. This detail is suitable for roofs up to 6 m in span. Also shown are two alternative means of forming the overhanging eaves.

▲ **Figure 2.78** Mansard roof details

Dormer windows

These are vertical windows that project from a sloping roof surface to provide daylight and ventilation to a room in the roof space. The triangular sides of the dormer (cheeks) above the main roof are framed out with studding and covered with tiles, sheet metal or timber cladding. The actual roof over the dormer window, which extends from the main roof, can be constructed in various ways, as illustrated in Figure 2.79, although flat or pitched roofs are most commonly used. Figure 2.80 is an isometric view of the basic framework construction for a flat roof dormer. Twin rafters are used on either side of the opening to carry the increased load. Figure 2.81 illustrates a section through a dormer window showing typical construction details.

▲ **Figure 2.79** Dormer roofs

▲ **Figure 2.80** Flat roof dormer window construction

▲ **Figure 2.81** Dormer details

Bolted roof trusses

The component parts of bolted trusses have overlapped joints, which are connected up using bolts and toothed plates. They are normally prefabricated in a shop and arrive on site in two easily managed halves, which must be bolted together once the truss has been positioned. Figure 2.82 shows a typical bolted truss with enlarged joint details. The trusses are spaced along the roof at about 1.8 m centres to support the roof components. The purlins and ceiling binders are then fixed to these. The spaces between the trusses are in-filled with normal common rafters and ceiling joists. Figure 2.83 is a line diagram of this arrangement.

Half-trusses are used in place of the crown rafter for hip-end roofs, normal hip and jack rafters being inserted as required. Timber roof trusses for industrial buildings are rarely encountered in new work as these are almost exclusively made of steel, although the type of trusses detailed in Figure 2.84 may be met in maintenance and rehabilitation work.

▲ **Figure 2.82** Bolted truss details

Ornamental roofs

This classification includes turrets, spires and domes. These are all forms of fairly small decorative roofs, normally weatherproofed in sheet lead, copper or zinc and used to cover small structures such as bell towers and church steeples, etc. The construction of each roof is basically the same irrespective of its plan shape, which may be square, polygonal or circular.

ridge

common rafters

purlins

roof trusses

ceiling joists

◄ **Figure 2.83** Roof arrangement using bolted trusses

equal

equal

equal

equal

rafter

tie

strut

bolted and connector joint

equal

equal

equal

◄ **Figure 2.84** Half elevation and plan of two alternative industrial roof trusses

sheet roofing

purlin

twin rafters

bolted and connector joint

packing

twin ties

twin

Figure 2.85 shows the plan and section of a hexagonal turret roof, which may be mounted on a brick or timber structure. It consists of a wall plate, halved together to form the plan shape. Hip rafters are birdsmouthed over the wall plate at their lower end and are tenoned into a final post at the top. The hips should be backed to enable the roof to be boarded. Jack rafters are used between the hips.

The main differences in the construction of turrets, spires and domes may be stated briefly. Spires tend to be taller and more steeply pitched than turrets. These therefore require special consideration in the method of tying down to the main structure because of the considerable wind pressures they are subjected to. Domes have exactly the same construction details as turrets except that the top edges of the hips and jacks will have been pre-cut to the required curve before erection, as shown in Figure 2.86.

▶Figure 2.85 Hexagonal turret roof details

�

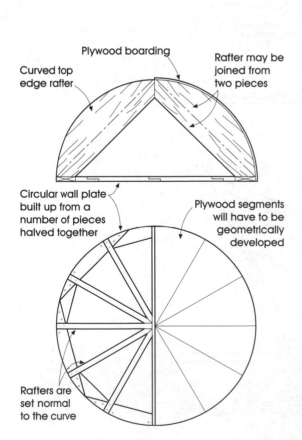

◀ Figure 2.86 Simple dome roof construction

Chapter Three

Partition walls

This chapter covers the work of the site carpenter and joiner. It is concerned with principles of partition wall construction. It includes the following:

- timber stud partition construction
- metal stud partition construction
- proprietary panel partition construction.

Partitions are internal walls used to divide space into a number of individual areas or rooms. Partition walls are normally of a non-structural, non-load-bearing nature. They are commonly formed from studwork framing, proprietary prefabricated systems or lightweight block work.

Timber stud partitions

Figure 3.1 shows the different types of partitioning. Studwork framing/partitions are commonly known as stud partitioning. This is a partition wall built of timber or metal studs, fixed between a sole and head plate, often incorporating noggins for stiffening and fixing (see Figure 3.2).

- *Stud.* A vertical timber or metal member of a partition wall fixed between the sole plate and head plate. The main member of a partition, it provides a fixing for the covering material.

Stud

Noggin

Sole plate

Timber studwork framing

Proprietary system

Lightweight blockwork

▲ **Figure 3.1** Partitions

▲ Figure 3.2 Stud partition components

▲ Figure 3.3 PAR and regularised (machined) sections have a constant width

- *Sole plate/head plate.* A horizontal timber fixed above or below studs to provide a fixing point for the studs and ensure an even distribution of loads.
- *Noggin.* A short horizontal piece of timber fixed between vertical studs of a partition. Its use stiffens the studs, provides an intermediate fixing point for the covering material and in addition a fixing point for heavy items which may be hung on the partition (WC cistern, hand basin, etc.).

Timber grade

Traditionally, timber is used for making studwork framing. PAR (planed all round) timber is preferred not only because of its uniform cross-section but also it is better to handle. Regularised or machined timber (timber machined to a consistent width by re-sawing or planing one or both edges) or ALS/CLS (American/Canadian Lumber Stock) processed timber with rounded corners may also be specified.

A consistent sized section aids the plumbing of a partition and also provides a flat fixing surface for the partition covering materials (see Figure 3.3).

Covering materials

The standard covering material for partition walls is **plasterboard**. Typical for stud partition use are thicknesses of 9.5 mm or 12.5 mm and sheet sizes of either 900 mm × 1800 mm or 1200 mm × 2400 mm. Tapered edge boards are best for a dry finish. These should be filled and taped to provide a smooth seamless joint that won't show through the subsequent paint or wallpaper finish. Square edge boards can also be used for a filled and taped dry finish, although they are primarily used for surfaces that are skimmed with a thin coat of wet board finish plaster (see Figure 3.4).

▲ Figure 3.4 Types of plasterboard

Jointing timber partition members

Traditionally, timber partitions were framed using basic joints to locate members and provide strength as shown in Figure 3.5. Studs were often housed into the head plate, slotted over a batten at the sole plate and mortised or tenoned at openings. However, present day techniques calling for speed of erection and economy result in the majority of partitions being simply butt jointed and skew nailed (Figure 3.6). An alternative to both methods would be the use of butt joints reinforced

Head plate

Housed into
head ⅓ depth

Stud

Slotted out to fit over batten

25 mm × 38 mm
batten

Sole plate

Wedge

Door head

Door stud

Sole plate

Dowel

◀ **Figure 3.5**
Traditional joint details

Butt jointed
and nailed

▲ **Figure 3.6** Nailed butt joints

Butt joints
and
framing
anchors

◀ **Figure 3.7** Use
of framing anchors

with metal framing anchors, as in Figure 3.7. These provide a quick yet strong fixing, although they are rarely specified.

Timber stud partitions may be constructed in situ or be pre-made and later erected.

- *In situ partition*. Components are cut, assembled and fixed in situation on site.
- *Pre-made partition*. A ready-assembled partition (either on site or in a factory) for later site erection.

Constructing in situ timber stud partitions

1. Mark the intended position of the partition on the ceiling. Ideally this should either be at right angles to the joists or positioned under a joist or double joist. Where the joists run in the same direction as the partition and it is not directly under a joist, noggins will have to be fixed between joists, to provide a fixing point (see Figure 3.8).

▲ **Figure 3.8** Fixing of head plate

2. Fix the head plate to ceiling, using 100-mm wire nails or oval nails at each joist position or 400-mm centres as appropriate. In older buildings, where ceilings might be in a poor position or easily damaged by nailing, 100-mm screws can be used as an alternative.

3. Plumb down from the head plate on one side at each end, to the floor, using either a straightedge and level or a plumb bob and line as shown in Figure 3.9. This establishes the position of the sole plate.

▶**Figure 3.9** Plumbing partition

4. Fix the sole plate in position with 100-mm wire nails, oval nails or screws as appropriate. Ideally, as with head plates, this should be either at right angles to the joists or over a joist or double joist. Where the joists run in the same direction as the partition but not directly under it, noggins will have to be fixed between joists to provide a fixing point, as shown in Figure 3.10. When the sole plate is fixed to a concrete floor, it should be plugged and screwed. Alternatively it may be possible to 'shot fix' both head and sole plates to concrete surfaces. Refer to your supervisor for permission/instruction.

5. Cut, position and fix the end wall studs, see Figure 3.11. These may be either:

▲ **Figure 3.10** Fixing of sole plate

▲ **Figure 3.11** Positioning wall stud

- plugged and screwed to brickwork and blockwork using 100-mm screws into proprietary plugs or twisted timber pallets;
- nailed to blockwork or into brickwork mortar joints using 100-mm cut nails;
- nailed directly to brickwork using 75-mm hardened steel masonry nails. (It is essential that eye protection is used when driving masonry nails as they are liable to shatter.)

6. Mark the positions of vertical studs on the sole and head plates. Studs will be required at each joint in the covering material and using 1200-mm wide sheets at 400-mm centres for 9.5-mm plasterboard or at 600-mm centres for 12.5-mm plasterboard. Using 900-mm wide sheets, all studs should be at 450-mm centres (see Figure 3.12). Similar centres may be employed for other covering material/cladding. Assuming 12.5 mm × 1200 mm covering, the second stud is fixed with its centre 600 mm from the wall, the third with its centre 1200 mm from the wall and the remaining studs at 600-mm centres thereafter.

◀ **Figure 3.12** Stud spacings

▲ **Figure 3.13** Skew nailing

7. Measure, cut and fix each stud to head and sole plates, using 100-mm wire or oval nails, driven at an angle (skew nailed, see Figure 3.13). Each stud should be measured and cut individually as the distance between the plates may vary along their length. Studs should be a tight fit: position one end, angle the stud and drive the

Marking stud length

▲ **Figure 3.14** Marking and positioning studs

other end until plumb. The length of each stud may be measured using either a tape or pinch rods as shown in Figure 3.14.

8. Mark noggin centre line positions. These will vary depending on the specification. Typically they are fixed at vulnerable positions where extra strength is required: at knee height 600 mm up from floor, at waist height 1200 mm up from floor and at shoulder height 1800 mm up from floor as shown in Figure 3.15. Where deep section skirting is to be fixed an extra noggin may be specified near its top edge for fixing purposes. Additional noggins may also be required where heavy items are to be hung on the wall, such as where a fixing is required for electrical mounting boxes and where the covering sheet material is jointed in the height of the partition (see Figure 3.16).

Additional tall skirting noggin

₵ of noggins

▶**Figure 3.15** Noggin positions

9. Cut and fix noggins (see Figure 3.17), either by skew nailing using 100-mm wire or oval nails or staggering either side of the centre line and through nailing again using 100-mm wire or oval nails.

Openings

Where door, serving hatch or borrowed light (internal glazing) openings are required in a studwork framing, studs and noggins should be positioned on each side to form the opening (Figures 3.18 and 3.19). The sole plate will require cutting out between the door studs. Linings are fixed around the framed opening to provide a finish and provision for hanging doors or glazing.

Additional noggins to provide fixing points

◀ **Figure 3.16** Additional noggins

Noggins staggered
Through nail

Noggins in line
Skew nail

Stud space template used when skew nailing

Noggin height template used for support when skew nailing

▲ **Figure 3.17** Fixing noggins

Door head noggin (transom)

Folding wedges

Door stop

Lining

Door stud

Sole plate cut away

Lining/frame width and height plus clearance

Door

▲ **Figure 3.18** Door opening in a stud partition and horizontal cross-section

▲ **Figure 3.19** Borrowed light opening in a stud partition and vertical cross-section

Returns

Should a return partition forming internal or external angles be required, stud positions must be arranged to suit. Consideration must be given to providing a support and fixing for the covering material around the return intersection. Two alternative methods, shown in Figures 3.20 and 3.21, are commonly employed.

▶**Figure 3.20** Corner details

The particular method adopted will depend on whether the carpenter is fixing the covering material as the work proceeds or the plasterer is fixing it after the carpentry work is complete. Sole and head plates in long partitions and those containing return intersections will require joining preferably using halving joints secured with screws (see Figure 3.22).

Cutting and fixing plasterboard

Plasterboard is normally marked by its manufacturer to indicate which face is suitable for plastering or dry decoration. Typically the following is marked on the rear face: 'Use other face for plastering and decoration'. In general the ivory face is fixed to the outside.

Extra stud or batten
to hold plasterboard

Corner

Extra stud
or batten

'Tee' junction

◀ **Figure 3.21** Stud treatment to provide support for covering at returns

▲ **Figure 3.22** Sole and head plate joints

Cutting

Plasterboard can be cut with a saw or with a replaceable blade craft knife (see Figure 3.23):

■ When sawing use the saw at a low angle to the sheet.

■ When using a knife, score a fairly deep line into the face along a straightedge. Apply downward pressure to snap the sheet along the scored line. Run the knife along the snapped line on the underside to cut the paper and separate the two pieces.

■ Holes for electrical fittings may be cut using a pad saw or with a craft knife.

Use hand saw at a low angle

Score line on face using knife against a straight edge

Snap board on scored line, use knife to separate

Cut holes using a pad saw

▲ **Figure 3.23** Cutting plasterboard

Fixing procedure

Plasterboard sheets are secured to the studs using plasterboard nails or plasterboard screws at approximately 150-mm centres around the sheet edges and intermediate studs. To avoid distortion, nailing should commence from the centre of the sheet working outwards. Nails should be driven just below the surface, but taking care not to break or damage the paper face of the sheet. See Figure 3.24 for the fixing procedure:

■ Pre-cut the sheets to length, making them 10 to 15 mm shorter than the floor to ceiling height.

■ Starting either from the corner or a doorway opening, use a board lifter to position and hold the board in place up against the ceiling; secure to all studwork members.

■ A narrow strip must be cut from the edge of the board over a doorway, to enable two boards to butt on the centre of the stud. Alternatively a short stud can be fixed to the side of the door stud to provide a fixing.

■ Position and secure the remaining full boards in place working across the position. Tapered edge boards can be lightly butted together, square edge boards should be fixed leaving about a 3-mm gap between them.

■ Cut final board to width, position and secure.

■ Repeat the procedure to cover the other side of the partition.

Where services are to be accommodated in the partition, they should be installed either before any plaster boarding is done or before the second face is boarded.

Fix first board in corner or at a door opening

Plasterboard cut back or short stud fitted to provide fixing for over door piece

Nail at about 150 mm centres to all studs and noggins

Use board lifter to hold board in place, tight to the ceiling

▶ **Figure 3.24** Fixing plasterboard

Pre-made partitions

These may be made either by the joiner in a workshop or by the carpenter on site. The method of jointing will normally be either: studs housed into plates with noggins butt-jointed and nailed, or all joints butted and nailed. In addition, occasionally pre-made partitions may be jointed using framing anchors.

This type of partition must be made undersize in both height and width, in order for it to be placed in position. Once in position folding wedges are used to take up the positioning tolerance prior to fixing (see Figures 3.25 and 3.26). Plates and wall studs are fixed using the same methods and centres as are applicable to in situ partitions.

▲ **Figure 3.25** Pre-made partitions must be constructed undersized to allow for positioning tolerances

▲ **Figure 3.26** Use of folding wedges to take up positioning tolerances

Drilling and notching partitions

Service cables and pipes for water, gas and electricity are often concealed within stud partitions. In common with floor joists the positioning of holes and notches in studs has an effect on strength. It is recommended that holes and notches in studs should be kept to a minimum and conform to the following, see Figure 3.27:

■ **Holes**. Up to one quarter (0.25) of the stud's width, drilled on the centre line (neutral stress line) and located between 25% and 40% (0.25 and 0.4) of the stud's height from either end are permissible. Adjacent holes should be separated by at least three times their diameter measured centre to centre.

■ **Notches**. Permitted on either edge of the stud up to 15% of the stud's width and located up to 20% (0.2) of the stud's height from either end.

▲ **Figure 3.27** Hole and notch details and positioning

Example

The position and sizes for holes and notches in 100 mm width studs, 2400 mm in length are:
- *Holes.* Between 600 mm and 960 mm in from either end of the stud and up to 25 mm in diameter.
- *Notches.* Up to 480 mm in from each end and up to 15 mm deep.

Excessive drilling and notching outside these permissible limits will weaken the stud and may lead to failure.

In addition, holes and notches should be kept clear of areas where the services routed through them are likely to get punctured by nails and screws, e.g. behind skirtings, dado rails and kitchen units etc., or metal plates can be fitted for protection.

Fire resistance, thermal insulation and sound insulation

Depending on the partition's location it may be required to form a fire resisting and/or a thermal insulating and/or a sound insulating function. See Figure 3.28. Specific requirements and methods of achieving them are controlled by the Building Regulations. These should be referred to for further information as may be required.

However, the general principles for achieving these functions are as follows.

Fire resistance

This can be increased by a double covering of plaster board, the second outer layer being fixed so that the joints overlap those in the lower layer.

Thermal insulation

This can be increased by filling the space between studs with either mineral wool or glass fibre quilt.

Sound insulation

This can be increased by a discontinuous construction to avoid impact and vibration and the use of lightweight infilling material such as mineral or glass fibre quilt to absorb sound energy.

▲ **Figure 3.28** Fire, heat and sound insulation of stud partition

Metal stud partitions

Various types of metal stud partition systems are available from a number of manufacturers. A typical system for domestic use consists of:

- galvanised 'U' section channels for fixing to the floor and ceiling;
- 'U' or 'H' sections for intermediate studs;
- 'U' section channels for wall studs and use at openings or junctions;
- 'U' section channels for noggins.

Typical sections and sizes are shown in Figure 3.29.

All members are pre-slotted to accommodate electrical cables and other service runs. Rubber grommets must be used where electrical cables pass through members in order to prevent chafing. Services are normally installed after one side of the partition has been covered.

▲ **Figure 3.29** Metal stud partition members

▲ **Figure 3.30** Metal stud junction details

Studs at 900 mm centres

Electric cables clipped
to rear of plasterboard

'U' wall stud

Rubber gromit used
where electric cables
pass through metal
sections

'U' noggin

Plasterboard fixed
with self-tapping
screws

Skirting block
(noggin off cuts)

'H' stud

'U' floor section

Acoustic mastic applied to
back sections at wall, floor
and ceiling junctions

▲ **Figure 3.31** Metal stud erection

A range of typical fixing details for various junctions is illustrated in Figure 3.30. Timber sole plates may be specified under the floor channel especially where uneven or out-of-level floors are encountered. 12.5 mm or 15 mm plasterboard is recommended for covering metal stud partitions. Fitting two layers of plasterboard with the joints staggered and filling the spaces between the studs with insulating material will improve the partition's thermal, sound and fire insulation properties.

Erection sequence

Figure 3.31 illustrates the erection sequence of metal stud partitions:

- Mark the position of the partition on the floor, ceiling and walls.
- Screw fix floor and ceiling channels and end wall studs at 600-mm centres. A continuous 6-mm bead of acoustic mastic/sealant applied to the back of all members before positioning will enhance the partition's sound insulating properties.
- Cut intermediate studs to length and insert studs between floor and ceiling channels at 900-mm centres.
- Insert skirting blocks centrally between each pair of studs. These may be a standard item or cut from a noggin section.
- Insert noggins at a height of 1200 mm.

Depending on the actual make of partition components being used, members may friction fit together, have interlocking lugs or may require fixing together with self-tapping screws:

- Cut, position and screw fix plasterboard to one side of the partition. Use self-tapping bugle-head plasterboard screws at up to 400 mm centres.
- Cut and fix plasterboard on the other side of the partition.

Door openings

At the head of door and other openings, channels will have to be fabricated to fit between the studs, see Figure 3.32:

- Cut a length of floor or ceiling channel to span the opening plus about 150 mm extra at each end.
- Cut through the side flanges at each end and bend at right angles.
- Use self-tapping screws to fix the door head channel.

▲ Figure 3.32 Fabricated door head

Proprietary panel partitions

Various proprietary partition systems are available from a number of manufacturers, but probably the most popular for domestic work is the cellular core dry partition system. This is constructed using prefabricated panels, which consist of two sheets of plasterboard separated by and bonded to a paper/cardboard cellular core, see Figure 3.33.

These are available in panel widths of 900 mm, and 1200 mm, thicknesses of 57 mm and 63 mm, and heights of 2350 mm, 2400 mm

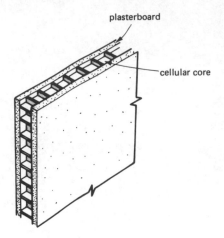

▲ Figure 3.33 Proprietary dry partition board

▲ Figure 3.34 Proprietary panel partition

and 2700 mm. The 57-mm thick panels use 9.5 mm plasterboard and the 63-mm boards use 12.5 mm.

The panels rest on timber sole plates fixed to the floor and fit over timber battens which are fixed to the wall and ceiling. Battens are also fixed between adjacent panels (see Figure 3.34). All battens that fit into the panels should be a press fit, and therefore may require some easing. A bevel may be applied to the leading edges of battens to facilitate the easy entry into the panel edge.

A range of typical fixing details for various junctions is illustrated in Figure 3.35.

Erection sequence

1. Mark the position of the partition on the floor, ceiling and walls.
2. Fix the panel width sole plate to floor along the total length of the partition.
3. Fix the 19 mm × 37 mm bevelled edge battens to the ceiling and end walls.
4. Fix 19 mm × 37 mm × 150 mm location blocks to the sole plate at the wall ends.
5. Use a handsaw to cut first panel to length; this should be 3 mm shorter than the distance between the top of the sole plate and ceiling.
6. Insert skirting fixing battens into the bottom edge of the panel, one for 900-mm panels and two for 1200-mm panels.
7. Locate the top of the panel over the ceiling batten about 200 mm from the wall; swing the panel to the vertical position, resting on the sole plate; slide the panel along the sole plate to locate over the location block and wall batten.
8. Insert 300-mm long location block half way into bottom corner of panel and fix to sole plate.
9. Cut to length and insert a 37 mm × 37 mm joint batten. Tap halfway into the edge of the panel.

▲ Figure 3.35 Proprietary partition details

10. Nail or screw fix the panel to all battens, using either 38-mm plasterboard nails at up to 150-mm centres or 38-mm bugle head plasterboard screws at up to 300-mm centres.

11. Cut and fix remaining panels using the previous procedure.

It is not normally necessary to rake out the panel core before inserting battens. Gently tapping/pushing the batten into the panel edge/end should force back sufficient core to accommodate the batten. Alternatively the core on the top and two long edges may be removed prior to positioning by raking out with the claw of a hammer.

Services

Electric cables and other similar services up to about 30 mm in diameter can be accommodated in the cellular core. Use a spiked rod to hollow out the core to the position of the electrical outlet before positioning the panel. The wires should be fed down the panel to the outlet box hole as the panel is offered up to the ceiling batten and slid into place.

Timber plugs (37 mm × 37 mm) can be inserted into the cellular core in various positions during erection to provide later fixing points for cisterns, radiators, washbasins, etc.

Door openings

At door and other openings the joint batten should be tapped into the panel until it is flush with the edge. Fix short lengths of 19-mm batten to the upper end of the joint battens at either side of the door head. Cut the door head panel from an offcut, slide into position and fix with nails or screws, see Figure 3.36.

Fixing of final panel

Where a partition is required between two walls and it does not contain a door opening or offset, a different procedure is required to position and fix the final panel as illustrated in Figure 3.37:

1. Use the previous erection sequence, but work from both end walls, making the final panel somewhere in mid run between two fixed panels.

2. Rip out the core on both long edges of the final panel to allow the joint batten to be inserted flush with the edges of the panel. Also rip out the core on the edges of both fixed panels to a depth of 19 mm.

3. Cut the two bottom corners off the final panel to allow it to fit over the sole plate location blocks. Also cut two 20-mm slots in both edges of the panel, about 450 mm from the top and bottom.

4. Place joint battens halfway into the edge of the final panel and partly insert a nail or screw into the battens at each slot position. Push the battens fully into the panel edges.

5. Position final panel over ceiling batten and swing into place. Use the protruding nails or screws to draw the joint battens into the edges of the two fixed panels.

6. Secure the panels to all battens and finally refit and fix the two cut-out pieces of panel at the sole plate.

Ceiling batten

Door head batten
fixed to door batten

Door head panel
pushed up into
position

Door batten

Door head
inserted in core

▶**Figure 3.36** Door head panel

Final panel

Joint batten
fully inserted
in both edges

Screws to hold
joint batten

Use the screws to draw
the batten centrally
into the fixed panel

Cut slots in
edge of panel

Corners removed to
fit over sole plate
location block

Re-fix the cut
out corners

▶**Figure 3.37** Fixing of final panel

Fire resistance, sound and thermal insulation

The properties of these partitions increase with the panel thickness.
The sound, thermal and fire properties of the partition may be further
improved by bonding an additional layer of plasterboard to one or
both sides with a suitable adhesive. For the greatest benefit the joints
between the panels and sheets should be staggered.

Chapter Four

Timber-framed construction and timber-engineered components

This chapter covers the work of the site carpenter and bench joiner. It is concerned with principles of timber-frame construction and timber-engineered components. It includes the following:

§ an introduction to types of timber-framed buildings
§ construction and erection of temporary timber-framed buildings
§ construction and erection of permanent timber-framed buildings
§ details of glulam components, stressed skin panels and built-up beams
§ construction and erection of shell roofs
§ details of curtain walling.

Timber-frame construction

Timber-framed construction can be divided into two main types:

■ *Temporary 'knock-down' buildings.* These are buildings, which remain in one position for a relatively short amount of time, and are 'knocked down' (dismantled) and re-erected elsewhere. Site huts, sheds, stores and offices can all fall within this group.

■ *Permanent buildings.* Mainly factory built using either platform or balloon frame methods of construction; used for domestic housing, office and industrial accommodation. Also includes timber-framed panels (curtain walling) and other structurally engineered timber components.

Terminology

■ *Balloon framing.* A form of timber-frame construction where the studs forming the wall panels are continuous from the ground level up to the eaves, with intermediate floor joists being supported on ribbons.

■ *Platform frame.* A form of timber-frame construction in which the walls are constructed in single storey height panels; floors extend over the wall panels at each level thus forming a platform on which the walls above can be erected.

- *Glulam.* Glue laminated timber, normally engineered structural timber members glued and layered up from fairly small, sections and lengths to form either large cross-sections, long lengths, curved shapes or a combination.

- *Stressed skin panel.* A structural unit consisting of timber framing members on which a plywood skin is fixed to one or both sides. The panel functions as a whole transmitting the stresses between the skin and framing. Used as prefabricated panels for floors, walls and roofs.

- *Built-up beams.* A timber engineered structural member, consisting of top and bottom flanges, spaced by either ties and struts or plywood webs and stiffeners.

- *Shell roof.* A timber shell roof consists of a thin curved lightweight membrane acting as a stressed skin that obtains its strength as a result of its curvature. Laminated edge beams are incorporated around the perimeter to carry the membrane's forces and strengthen the edge of the shell.

- *Curtain wall.* A non-load bearing infill wall unit, consisting of timber framed panels and windows, used as the external enclosing wall for skeleton frame and cross wall construction.

Temporary 'knock-down' buildings

The use of each building will determine its size and the standard of facilities required. In general, site accommodation needs to be well ventilated, waterproof, thermally insulated and have some provision for lighting and heating.

A typical **site hut** is illustrated in Figure 4.1. The framing sections for the walls, roof and floors which bolt together are all made up in standard sizes, thereby making the panels fully interchangeable and enabling larger or smaller huts to be erected as required. In order to facilitate this, all bolt holes should be drilled using a template. Where sheet material is being used the ideal panel width is 1220 mm. The larger window or door panels should be made up in multiples of the panel size, e.g. 2440 mm for sheet material.

The wall covering material (cladding) may consist of either shiplap boarding, matchboarding or WBP plywood. Where plywood is used for the cladding, the diagonal braces can be omitted, as the sheets themselves will provide sufficient rigidity. Where shiplap or matchboarding is to be used for the cladding, a moisture barrier can be fixed between the cladding and the studs to prevent the penetration of wind-driven rain. Plywood or tongue and groove boarding can be used for the floor and roof coverings.

The hut should be erected on a level base of sleepers and to prevent dampness and decay, a damp proof course can be inserted between the sleepers and the floor panels/joists.

▲ **Figure 4.1** Site hut

Figure 4.2 shows a section through the wall panel showing the opening casement. Fixed glazing could be used as an alternative if required, but provision must be made for ventilation. This could take the form of a timber louvred ventilator placed over the door or in a wall panel (see Figure 4.3).

▲ **Figure 4.2** Window detail   ▲ **Figure 4.3** Ventilator detail   ▲ **Figure 4.4** Lining and insulation

The inside of some site huts are lined with insulation board. This not only improves their thermal insulation but also their internal appearance. Fibreglass or mineral wool insulation is placed between the studs to provide additional insulation (see Figure 4.4). Where the hut is lined, a batten must be fixed to the corner panel to take the lining and pockets must be left for the bolts.

Permanent timber-framed buildings

In the UK, permanent buildings are normally prefabricated under factory conditions and transported to the site for erection on prepared foundations. In other parts the 'stick built' method is more commonly utilised. This is where the timber is delivered to the site as separate loose timbers and framed up adjacent to where it will be erected subsequently.

In timber-framed buildings all of the structural parts above the damp-proof course, that is the walls, floor and roof, are constructed from timber and plywood. The external cladding merely provides a weatherproof finish and may include bricks, stone, tiles, cement rendering and various types of timber boarding.

Timber-framed construction methods have a number of advantages over other forms of construction. These include:

- Reduction in the site erection time.
- Improved standards of thermal insulation.
- Wide variety of external finishes available.
- Dry construction methods allow quicker occupation as no drying out time is required.
- No subsequent remedial work as a result of shrinkage is necessary.
- Adverse weather conditions do not greatly affect the construction progress.
- Prefabricated components are manufactured under ideal factory conditions.

The two principal designs of timber frame units are illustrated in Figure 4.5, these being:

- Platform frame (small panel, large panel and volumetric)
- Balloon frame (large panel)

Platform frame

This is by far the most used method of timber frame construction. The walls are built in single-storey sections. Once the first wall sections have been erected on the ground floor, the first-floor joist and floor covering are fixed on top of them. This creates a platform on which the second-storey walls can be erected. Further floors are formed by continuing this process. Finally the roof is erected on top of the upper wall panels. A variation of the platform frame method, which utilises

▲ **Figure 4.5** Timber frame units

factory construction to a much larger extent, is volumetric housing units. In this, complete box or room-size units are delivered to the site and stacked together to form the completed structure. These are normally fully decorated and have all services installed, thus reducing the on-site construction time to an absolute minimum. They have the disadvantage of requiring the use of a large, powerful crane.

Balloon frame

This method uses timber frames that are the full height of the building. The studs are continuous from damp-proof course level to the eaves. The wall panels are therefore erected to their full height in one operation, with intermediate floors being supported on ribbons notched into the continuous studs.

Methods of construction

The platform frame method is normally preferred because of its smaller, more easily managed components and the fact that a working platform is erected at each stage. The panels, which comprise head and sole plates, studs and sheathing, are normally related to the maximum size and weight that can be handled by two people. This gives panel sizes of single-storey height by approximately 3.600 m. Unsheathed panels for internal partitions may be considerably larger. In situations where a crane is available on site to erect the superstructure, the maximum panel size restriction is removed. This enables the use of balloon frames and larger platform frames. This method is often known as the large-panel erection system. It allows full house width and depth panels to be used; in addition, these panels are often more highly factory finished, including windows, glazing, insulation, internal lining and, where appropriate, external cladding. This has the advantage of reducing the on-site time and enables more rapid completion.

External walls

Typical examples of the external wall construction using both platform and balloon frame methods are shown in Figure 4.6. The majority of timber frame buildings are constructed with solid concrete ground floors. Where a suspended timber ground floor is required, the detail shown in Figure 4.6 (c) may be used.

Upper floors

The upper floors of timber frame buildings are normally simply constructed on site using loose joists and boarded over with either tongued-and-grooved boarding or sheet material. However, where a contractor is using a crane for large panel erection, the floor panels may be prefabricated in large sections, using either standard joists and sheet material or specially designed stressed skin panels.

Partitions

The internal partitions of a timber frame building are normally erected along with the main structure. They are constructed in exactly the same way as the external wall panels, including lintels over door openings, etc. The only difference is that they will have no structural sheathing prefixed. This tends to make them unstable when handled. Therefore it is normal to fix a temporary diagonal brace across the face of the studs to stiffen and keep the panels square during erection.

Roofs

The roofs of timber frame buildings are the same as those used for other methods of construction and therefore the roof details covered in the previous chapter equally apply.

Erection sequence

The sequence of operations for the erection of both methods of timber frame construction is similar. Once the foundations are complete the major components may be delivered from the factory. These can then be erected in one or two days to provide a fully weatherproof structure.

▲ Figure 4.6 Platform frame (a), balloon frame (b) and suspended ground floor (c)

▲ **Figure 4.7** Components and erection sequence for timber frame buildings

Finishing work can be carried out inside at the same time as the external cladding is being undertaken.

Figure 4.7 illustrates the major components and erection sequences for both methods of construction.

Care must be taken during erection, as until the shell is fully completed, including floor decking, roof coverings and glazing, the timber frame may be structurally unstable. This creates an obvious risk during the erection; to a great extent this risk may be avoided by the use of temporary diagonal braces and struts, particularly in windy conditions.

Fixings to foundations

The first item to be fixed is the lower sole. Its purpose is to form a level base on which subsequent components can be erected and also to transmit vertical and wind loads down to the foundations. As the lower sole provides an important link between the substructure and the superstructure, care taken in accurately positioning and fixing it will prevent any inaccuracies being repeated throughout the structure.

Illustrated in Figure 4.8 are two alternative methods of fixing the sole plate, either by nailing with a cartridge-operated fixing tool or by using some form of rag bolt or expanding bolt. It can be clearly seen that the use of either method will hole the DPC. With nails this problem is not considered significant as the damp-proof course tends to self-seal around the tight perforation made by the nail, and as the sole plates are pressure impregnated with preservative any minute amount of moisture will have no harmful effect. Where bolts are used, any gap between the damp-proof course and bolt can be sealed with mastic or a fibre washer.

▲ **Figure 4.8** Sole plate fixing

The fixing down of the sole plate, the nailing of the panels to it and the nailing of adjacent panels to each other are a critical part of the structural design. Each timber frame designer will

▲ **Figure 4.9** Typical nailing schedule

therefore produce a comprehensive **nailing schedule**, which must be
fully complied with both during manufacture and on site. A typical
nailing schedule is illustrated in Figure 4.9.

External wall panels

These normally consist of prepared 50 mm × 100 mm strength-graded
softwood studwork. The studs, which are spaced at 400-mm or
600-mm centres to suit the sheet sizes, are simply butt jointed and
nailed to the head and sole plates. No nogging or diagonal braces are
required as the plywood sheathing, which is nailed to the studs,
provides the required rigidity. An additional head plate is used to bind
the panels together, line them and provide additional strength, so joist
and trussed rafter spacing are not dependent on the stud positions.
Where wall panels intersect additional studs should be used to provide
a fixing for the internal lining (see Figure 4.10). It should be noted that
the head plate joints are staggered at wall panel intersections to tie
them together.

Lintels must be provided over openings where these occur in external
walls. This is normally two pieces of 50 mm × 150 mm timber fixed
together and stood on edge. This is supported on either side of the
opening by double studs, one stud being cut shorter (cripple stud) to
provide a bearing for the lintel (see Figure 4.11). Where wall panels
contain a door opening it is advisable to leave the sole plate in position
during the erection process to avoid panel distortion and damage.

double head plates

double head plates

studs

packing piece

studs

packing pieces

external corner

tee junction

▲ **Figure 4.10** Corner details

two pieces
to form
lintel

cripple
stud to
support
lintel

upper sole plate
at door openings
removed after
erection

▲ **Figure 4.11** Opening details

On-site construction (stick built)

Although in general the majority of timber frame buildings are produced in factory conditions and some even by computer-controlled cutting and assembly machines, timber frame is equally suited to the on-site stick-built method. This is especially true of the one-off design constructed by a small builder. All of the previous details will still apply to the on-site method, and if care is taken in the making of the panels the same accuracy in sizes as achieved by the factory method is possible, e.g. +0 mm and –3 mm in both directions.

Figure 4.12 illustrates a jig suitable for the on-site manufacture of timber frame panels. It consists of a plywood deck on timber bearers, supported by trestles. Fixed stops and wedges are used to locate the member and cramp up the panel.

window sill
height marked
on table

fixed stops
offset to
enable end
nailing

folding wedges
to cramp up
panel

fixed stops
to locate studs

▶**Figure 4.12** Jig for on-site
panel making

114

Preservative treatment

All external wall timbers, sole plates, battens, plywood, cavity barriers and timber cladding should be preservative treated to protect them against decay and insect attack, preferably using a vacuum or pressure treatment. All timbers cut after treatment, whether in the factory or on-site, must be given two flood brush applications of the preservative on the cut surfaces.

Barriers

Two essential barriers must be incorporated in the construction. These are a vapour barrier and a moisture barrier.

Vapour barrier

It is most important that a vapour barrier is used on the warm side of the insulation. Its purpose is to minimise the passage of water vapour from the moist interior of the building passing into the structure and condensing (interstitial condensation). If this were allowed to take place the wall materials would become saturated, thus making the insulation ineffective and creating the ideal conditions for the development of timber decay.

The vapour barrier, which is normally of 250–500 gauge polythene sheet, should be situated between the insulation and plasterboard. It should be fixed to the head and sole plates, around openings and overlaps with staples at approximately 250 mm. All joints should be overlapped by at least 100 mm and occur over a stud.

Care should be taken in fixing the polythene to ensure that it covers the entire wall surface, including the sole plate, head plate, studs, lintels and reveals. Any holes around services or tears in the barrier must be repaired with a suitable adhesive tape to seal any gaps before the plasterboard is fixed.

So as not to seal moisture within the wall structure, it is most important that the moisture content of the framework is checked prior to the fixing of the vapour barrier. This should always be less than 20%.

It is impossible for a vapour barrier to be 100% effective because of the joints in it and fixings through it. Therefore a breather-type moisture barrier is installed on the outside of the timber frame under the cladding so that any water vapour that has penetrated through the vapour barrier is allowed to escape or 'breathe' to the outside air.

Moisture barrier

A moisture barrier must be waterproof but at the same time must allow water vapour to pass through it. Fixed on top of the sheathing, it 'weathers' the timber frame shell immediately after erection. In addition it also provides a second line of protection against the penetration of wind-driven moisture that might penetrate the timber cladding.

Moisture barriers also seal any gaps in the panel joints and thus act as a draughtproofer.

Moisture barriers are usually special lightweight building paper or felt, designed to be both waterproof and vapour permeable (able to breathe). They are fixed to the sheathing starting at the bottom of the wall in horizontal widths around the building. They should be fixed to the head and sole plates, studs, around openings and overlaps with staples at approximately 250-mm centres. Horizontal laps should be at least 100 mm, with the upper layer overlapping the lower one. Vertical laps should be at least 150 mm and staggered in adjacent layers (see Figure 4.13).

▶**Figure 4.13** Wall panel insulation and barriers

Staples may be driven through narrow plastic strips. This avoids the barrier pulling over the staples, particularly when exposed to windy conditions. In addition the strips also serve to indicate the stud positions for subsequent wall tie fixing.

Insulation

In accordance with the Building Regulations, all buildings and extensions must be thermally insulated to a high standard. This is easily achieved in timber frame buildings by installing typically fibreglass or mineral wool insulation in the spaces between the studs of the external walls and between the ceiling joists in the roof space. Suspended ground floors may be insulated between the joists with fibreglass or mineral wool insulation supported on galvanised wire or plastic mesh fixed to the underside of the joists. Solid ground floors can be insulated by placing a suitable material, e.g. 50-mm thick polystyrene sheet, between the floor slab and screed.

Fire stops and cavity barriers

The Building Regulations require fire stops and cavity barriers to be fitted in the timber frame construction to control the spread of flame or smoke through cavities.

A fire stop is a non-combustible material used to divide the separating wall cavities of semi-detached or terraced buildings from the external wall or roof cavity (see Figure 4.14). In addition, fire stops are also required where services pass through the timber structure.

Timbers of at least 38 mm section are often used as cavity barriers. They must be fixed to seal the cavity between the timber frame and cladding, horizontally at each floor and roof level, around all door and window openings and vertically at intervals of not more than 8 m. In addition, vertical cavity barriers are required to separate the external wall cavities of semidetached or terraced buildings (see Figure 4.15 for details).

▲ **Figure 4.14** Fire stop and cavity barriers

◀ **Figure 4.15** Cavity barriers

Figure 4.16 illustrates how cavity barriers should be protected from the external wall by a damp-proof course.

◀ **Figure 4.16** Protecting cavity barriers

Cladding

Various types and arrangements of material can be used for cladding the exterior of timber frame buildings. The cladding provides a visually acceptable and sufficiently durable weatherproof finish. The main types of cladding in common use are illustrated in Figure 4.17.

The most popular cladding is brick veneer, which gives the building a traditional appearance, although this is often used in conjunction with large areas of timber cladding, tile hanging or cement rendering.

Brick cladding

This must be effectively connected to the timber frame for stability. Since there will be differing amounts of movement between the timber frame and the brick cladding, flexible wall ties must be used.

▲ **Figure 4.17** Main types of cladding

In addition, gaps should be left between the brickwork and the timber to allow for movement at the eaves and around window and doorframes that are fixed to the timber frame. These gaps around the window and doorframes should be sealed with compressible mastic to prevent rain penetration.

The cavity between the brick cladding and the timber frame should be ventilated at the eaves and below the damp-proof course level, by raking out the vertical brick joints (perpends) at 1500-mm intervals. The lower open joints also serve as weep-holes to drain the cavity.

Other lightweight cladding

Timber, tile hanging and cement rendering are preferably fixed to vertical battens nailed to the studs or horizontal and vertical counter-battening. This allows for free movement of air so that the cladding is ventilated on both sides. When timber boards are used, care must be taken when nailing to ensure that the boarding is nailed near one edge only and that the nails do not penetrate two boards. This is to ensure that each board is able to move more freely, without danger of splitting, when subsequent moisture movement takes place. Horizontal timber boarding is nailed near the bottom edge of the board. The top edge is held by the overlapping edge of the board above.

The length of nails used to fix the cladding must be $2\frac{1}{2}$ times the cladding thickness and should be treated to resist corrosion or be made of a non-corrosive metal.

Naturally durable timber cladding such as Western red cedar may be used without preservative treatment or any subsequent finish. Other timbers that are not naturally durable must be preservative treated. They may also have a further finish applied for decorative reasons or to prevent natural fading.

Paints and varnishes are the commonest finish. However, they are not particularly suitable as cracks in the film soon appear along the large number of joints in the cladding. This allows moisture to penetrate, causing the paint or varnish to peel.

Exterior wood stains are available in a wide range of colours. These are more suitable than paints or varnishes as they give the cladding a water-repellent surface without forming a film, and therefore cannot crack and peel off. Stains may be applied by brush, or by pressure impregnation at the same time as the preservative treatment.

Cladding details

Careful detailing to allow for differing movement and prevent rain penetration is required where claddings finish around door and window openings or where different claddings join. Figure 4.18 illustrates a number of typical cladding details.

▲ **Figure 4.18** Typical cladding finish details

Stressed skin panels

A stressed skin plywood panel consists of timber framing members called webs, onto which a skin of plywood is fixed. The panels can be single skin for most situations or double skin for use in high-load conditions. A double-skin panel is illustrated in Figure 4.19. The panels are used as prefabricated floor or roof panels, especially where long uninterrupted spans are required.

Stressed skin panels differ from normal joists and boarding in that the plywood, in addition to supporting the load between the framing, also contributes to the strength and stiffness of the framing itself. This is because the plywood is glued and nailed to the framing. Thus the panels function as an efficient structural unit transmitting the stresses between the plywood and framing.

▶**Figure 4.19** Stressed skin panel

Glue-laminated timber (glulam)

Structural timber members, which are of a large cross-section, long in length, or shaped, can be fabricated by the process of glue lamination (glulam). Examples of this are shown in Figure 4.20.

A glulam member comprises boards of small cross-section, glued and layered up either horizontally or vertically. The boards are layered with a parallel grain direction and cramped up in special jigs which are set up to provide the straight or curved profile of the finished member. A reduction in sectional size or greater span over solid beams of the same section is possible when using glue-laminated sections.

There are a number of reasons for this, in addition to the grade of timber used, the main ones being:

- The strength of timber is variable, but by joining a number of timbers together to act as one beam, the average strength of each layer (laminate) will be greater than the strength of the weakest laminate.
- Strength-reducing defects, such as large knots or shakes, are restricted in size to the thickness of one laminate.
- Higher-quality timber can be used in critical areas.
- Members can be pre-cambered to offset any deflection under loaded conditions.
- Because of the relatively small cross-section of each laminate, a tightly controlled moisture content can be achieved, resulting in a dimensionally stable finished member.

▲ **Figure 4.20** Examples of glulam

Site erection

Glulam members are often delivered on site shrink-wrapped in a plastic or waterproof covering. This should be left in place after erection to protect the member from the effects of weather and subsequent wet and finishing trades. Where members are to be jointed or fitted into metal shoes, etc. the protective covering can be temporarily pulled back.

The site work for glulam members normally involves jointing, erection, plumbing and fixing. In order to prevent accidental collapse, temporary ties and braces must be used until the structure is stable. Care must be taken to avoid distortions when lifting the members. For large structures manufacturers will provide or specify lifting points in certain positions to ensure that any stresses in erection do not exceed the permissible stresses in any member.

A range of typical joint and fixing details are illustrated in Figure 4.21. The main method of securing a member is with bolts and metal plates. The feet of arched or portal frames and posts are fixed to the base by the use of a metal shoe. This method has the added advantage of preventing moisture rising up the timber.

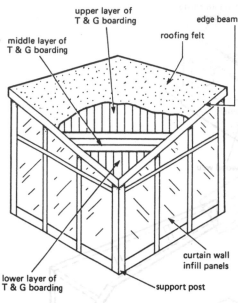

▲ **Figure 4.25** Hyperbolic paraboloid shell roof

upper layer of T & G boarding

middle layer of T & G boarding

edge beam

roofing felt

curtain wall infill panels

support post

lower layer of T & G boarding

produced between its diagonally opposite corners. Between the low corners a convex curve is produced and between the high corners a concave curve, whereas all cross sections parallel to the edges of the shell will be straight lines.

In general, shells with a greater curvature, i.e., those with a greater rise between high and low corners, are stiffer than shells of the same plan size but with a lesser curvature, i.e., a smaller rise between high and low corners.

Three layers of tongued-and-grooved boards glued with a synthetic resin adhesive and nailed together with square twisted-shank or annular nails form the hyperbolic paraboloid timber shell roof shown in Figure 4.25. Scaffolding is erected before construction to allow access to the underside of the roof, and more importantly to provide support for the lower layer of boarding. The middle layer is nailed to the lower layer, while the nails in the upper should penetrate right through the middle layer and into the lower layer. It can be seen that these boards are laid diagonally across the roof with the grain direction at right angles to its adjacent layers.

Any end jointing of the boards in the upper layers may be simple butts, although for strength it is normally recommended that the bottom layer is finger jointed. It is also advisable to use a better-quality timber on the bottom layer as this will be varnished to provide the finished ceiling.

Two layers of boarding may be used for the membrane of limited span shell roofs.

The edge beams, normally glulam, can be either made in one piece and positioned below the boarding or made in two pieces and positioned half above and half below the boarding as shown in Figure 4.26.

In order to stop the shell flattening out, the low corners are tied together with the aid of a mild steel tie bar that has been threaded at each end. Figure 4.27 illustrates this tie bar and also the method of fixing the edge beams in position.

Alternatively some form of buttress can be used to withstand the outward diagonal thrust of the shell at its low corners.

▲ **Figure 4.26** Shell roof edge beams

Wherever possible the edge beams should be given support by posts, walls or curtain walling mullions. In any case the low corners must be well supported as these take the main load of the roof. Also the high corners must be tied down to prevent lifting by wind forces.

Erection sequence

The sequence of operations for the erection of a hyperbolic paraboloid shell roof is as follows:

1. Erect corner support posts. Assemble scaffolding to support the first layer of boards.

2. Position lower half of beams and bolt to corner support posts.

3. Glue and nail each layer of boarding in position at right angles to each other, taking care to ensure that a close contact exists between all meeting surfaces.

4. Position upper half of beams. These are glued and fixed using either long square twisted-shank nails or coach screws.

5. Fix angle fillets to facilitate the dressing of the roofing felt over the edge beam. See Figure 4.28 for a typical edge detail.

6. Fix fascia.

7. Dismantle scaffolding.

8. Felt over the roof to make it watertight.

Alternative forms of shell

Multiple hyperbolic paraboloid shell roofs can be formed by joining a number of shells side by side (see Figure 4.29). Many other forms of shell roof are possible, some of which are also shown in the figure. Although vastly different in shape, these shells have a similar formation in that they all consist of a curved layered membrane.

▲ **Figure 4.27** Shell roof tie bar

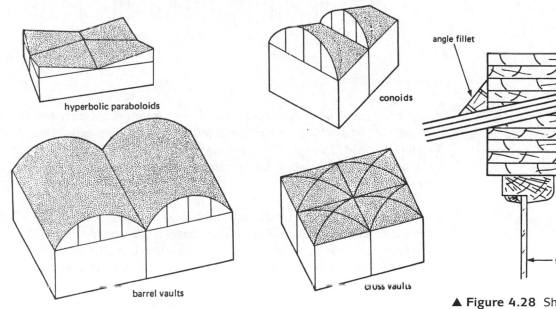

▲ **Figure 4.29** Shell roof forms

▲ **Figure 4.28** Shell roof edge detail

Curtain walling

Timber framed panels can be used as non-load-bearing in-fill wall units (curtain walling) for skeleton frame structures of either steel or reinforced concrete and for brick or concrete cross-wall construction.

These panels may be constructed as large window frames using a combination of glazed openings and insulated in-fill sections as illustrated in Figure 4.30. Typical vertical and horizontal details are shown in Figure 4.31.

▶**Figure 4.30** Curtain walling

The curtain walling frames are broken down into units bolted together at the head and sill. These horizontal joints incorporate a galvanised water bar bedded in mastic to produce an effective seal. A similar detail can be used to join frames side by side in order to form a long run of curtain walling.

Standard in-fill panels extend from head height of one frame to the normal windowsill height of the frame above. An alternative to the insulated timber clad panel shown, is the insulated, enamelled steel or GRP-faced panels used in some proprietary curtain walling systems.

Ventilation is provided by metal pivot-hung windows which are bedded in mastic and screwed to the main frame. The direct fixed glazing shows the use of both internal and external beads. The external ones should be permanently fixed with a suitable synthetic resin adhesive and pins. The internal beads are recess-cupped and screwed in place, thus permitting the glazing to be carried out from the inside of the building.

The panels must be securely fixed to the structural frame or cross-walls, normally using metal brackets, straps, or expanding bolts.

Careful detailing is required in order to allow for the differential movement between the curtain walling and the main structure while at the same time maintaining a weatherproof joint. This can be achieved by filling the joint with a resilient strip and sealing with compressible mastic.

resilient packing strip

jamb

glazed detail

mullion

mastic seal

jamb

spandrel detail

horizontal

metal pivot

sill

insulated spandrel panel

sill

galvanized bar bedded in mastic

head

bolt

direct glazing

vertical

▲ **Figure 4.31** Curtain walling sections

5

Chapter Five

Stairs

This chapter covers the work of the site carpenter and bench joiner. It is concerned with principles of stair construction. It includes the following:

¶ an introduction to types of stairs
¶ stairway construction and assembly
¶ installation of stairs
¶ construction of landings and railings
¶ protection of completed work.

A stairway can be defined as a series of steps (combination of tread and riser) giving floor-to-floor access. Each continuous set of steps is called a flight. Landings are introduced between floor levels either to break up a long flight, thus giving a rest point, or to change the direction of the stair where there is a restricted access.

Stair types

Stairs can be classified according to their plan shape:
■ straight-flight stair
■ quarter-turn stair
■ half-turn stair
■ geometrical stair.

In addition, each type can be further classified by its method of construction:
■ close string
■ open string
■ open riser
■ alternating tread
■ carriage beam
■ spine beam.

Straight-flight stairs

These run in one direction for the entire length. Figure 5.1 shows three different variations.

- The flight, which is closed between two walls, also known as a cottage stair, is the simplest and most economical to make. Its handrail is usually a simple section fixed either directly on to the wall or on brackets.

- The flight fixed against one wall is said to be open one side. This open or outer string is normally terminated and supported at either end by a newel post. A balustrade must be fixed to this side to provide protection. The infilling of this can be either open or closed and is usually capped by a handrail. Where the width of the flight exceeds one metre in width a wall handrail will also be required.

- Where the flight is free standing, neither side being against a wall, it is said to be open both sides. The open sides are treated in the same way as for the flight open one side.

closed between two walls

against wall open one side

free-standing open both sides

▲ **Figure 5.1** Straight-flight stairs

Quarter-turn stairs

As its name suggests, this type of stair changes direction 90° to the left or right by means of either a quarter-space landing or tapered steps (see Figure 5.2).

Tapered steps, although economising on space because of the introduction of the extra steps in place of the landing, are potentially dangerous owing to the narrowness of the treads on the inside of the turn. They should therefore be avoided where at all possible, especially in situations where they are likely to be used by young children or elderly persons, or at least be located at the bottom rather than at the top of the flight.

quarter space landing

quarter space of tapered steps

▲ **Figure 5.2** Quarter-turn stairs

Half-turn stairs

This stair reverses its direction through 180° normally by a half-space landing. Figure 5.3 illustrates the two main types.

- The dog-leg stair, where the outer strings of the upper and lower flights are joined into a common newel immediately above each other. This stair takes its name from the appearance of its sectional elevation.

half space landings

quarter space landings

dog-leg stair

open newel stairs

▲ **Figure 5.3** Half-turn stairs

■ The open-newel stair. This is also known as an open-well stair. Two newels are used at landing level. This separates the string of the upper and lower flights and thus creating a central space or well. A short flight is often introduced between the two newels at landing level, thereby creating two quarter-space landings and at the same time economising on space.

In addition, tapered steps could also be used instead of landings to negotiate turns in both dog-leg and open-newel stairs.

Geometrical stairs

The previous more robust types of stair utilise newels to change direction and also to terminate and support the outer strings. This gives them an image of strength and rigidity. A geometrical stair, as shown in Figure 5.4, presents a graceful, more aesthetic appearance. It has an outer string and a handrail that are continuous from one flight to another throughout the entire stairway. As both the string and handrail rise to suit the stairs throughout the curve they are said to be 'wreathed'.

Although the use of newels is not essential they are sometimes used at the top and bottom of a stair to support the scrolled handrail above. This type of stair is the most expensive and is thus rarely encountered in new work, except in 'one-off' high-quality constructions.

In addition to the stairs illustrated, an infinite variety of design for geometrical stairs may be devised, including helical (often misnamed 'spiral') and elliptical stairs.

Stair terminology

See Figures 5.5 and 5.6 for stair terminology

■ *Apron lining*. The boards used to finish the edge of a trimmed opening in the floor.

wall string may also be 'wreathed'

'wreathed' string and handrail

▲ **Figure 5.4** Geometric stair

Newel post
Handrail
Nosing
Wall string
Baluster
Upper floor or landing
Apron lining
String capping
Balustrade
Spandrel framing
Riser
Tread
Outer string
Bull nose step

▶**Figure 5.5** Stair terminology, close-string

130

▲ **Figure 5.6** Stair terminology, open riser and cut string

■ *Balustrade*. The handrail and the infilling between it and the string, landing or floor. This can be called either an open or closed balustrade, depending on the infilling.

■ *Baluster*. The short vertical infilling members of an open balustrade.

■ *Bull nose step*. The quarter-rounded end step at the bottom of a flight of stairs.

■ *Carriage*. This is a raking timber fixed under wide stairs to support the centre of the treads and risers. Brackets are fixed to the side of the carriage to provide further support across the width of the treads.

■ *Commode step*. A step with a curved tread and riser normally occurring at the bottom of a flight.

■ *Curtail step*. The half-round or scroll-end step at the bottom of a flight.

■ *Newel*. The large sectioned vertical member at each end of the string. Where an upper newel does not continue down to the floor level below it is known as a *pendant* or drop newel.

■ *Nosing*. The front edge of a tread or the finish to the floorboards around a stairwell opening.

■ *Riser*. The vertical member of a step.

■ *Spandrel*. The triangular area formed under the stairs. This can be left open or closed in with spandrel framing to form a cupboard.

- *String.* The board into which the treads and risers are housed or cut. They are also named according to their type, for example, wall string, outer string, close string, cut string, and wreathed string.
- *Tread.* The horizontal member of a step. It can be called a parallel tread, an alternating tread or a tapered tread, etc., depending on its shape.

Stairway construction and assembly

Stair regulations

Both the design and construction of stairs are very closely controlled by the Building Regulations. These lay down different requirements for stairs depending on the use of the building. These requirements are summarised and illustrated in Table 5.1.

Close string stair construction

The construction of any staircase with close strings follows the same basic procedure, with slight variations depending on the particular type of stair:

- The housings in the string can be cut out on a spindle moulder or by using a portable router and a stair housing template (see Figure 5.7). Where neither of these is available, they can be cut by hand using the sequence of operations shown in Figure 5.8.
- Bore out at nosing end with a brace and bit. Clean out nosing and cut edges of housing with a tenon saw.
- Clean out waste with chisel and hand router.

Template position lines

Batten fixed to bench top

Floor line

Plywood template

String

▲ **Figure 5.7** Portable router stair housing template

Table 5.1 Stair requirements (Building Regulations)

Requirement	Type of building where stairs are located		
	Private: Stairs for domestic use in a dwelling occupied by a single household	**Institutional and assembly:** Stairs serving places where a substantial number of people gather	**Other:** Stairs serving all other buildings
Pitch	Maximum of 42°	Governed by tread and riser requirements	
Number of risers	Good practice not to exceed 16 without an intermediate landing. Consecutive flights of more than 36 risers must change direction between flights by at least 30°	Maximum of 16 for assembly buildings. Consecutive flights of more than 36 risers must change direction between flights by at least 30°	Maximum of 16 for shops. Consecutive flights of more than 36 risers must change direction between flights by at least 30°
Rise of step	Maximum of 220 mm	Maximum of 180 mm	Maximum of 190 mm
Going of step	Minimum of 220 mm (Note: maximum rise cannot be used with the minimum going as it will be over the pitch requirement)	Minimum of 280 mm. May be reduced to 250 mm if the floor area served by the stair is less than 100 m².	Minimum of 250 mm
Combined rise and going	Twice the rise plus the going 2R + G to fall between 550 mm and 700 mm.		
Width of stair	No minimum requirement. Except when used as a means of escape, then the minimum width is determined by the number of people assessed as using the stair in an emergency: up to 50 people minimum width of 800 mm; 51 to 150 people minimum width of 1000 mm; 151 to 220 people minimum width of 1100 mm. Minimum widths for over 220 people depend on wether there is to be a phased or simultaneous evacuation (see Building Regulations for calculation formula). In all cases stairs for the disabled where there is not a lift, must be a minimum of 1000 mm between handrails.		

Pitch (diagram labelled "Pitch")

Number of risers (diagram): Landing, Landing, Up, Max 16 risers per flight, Max 36 risers before change of direction, At least 30°

Rise of step (diagram): Rise, Top of tread to top of tread

Going of step (diagram): Going, Nosing to nosing, Going, Face of riser to face of riser

Combined rise and going (diagram):
R = 200 mm
G = 275 mm
2R+G = 675 mm
OK for private use

Width of stair (diagram): Wall to handrail, Handrail to handrail, Wall to handrail on balustrade, Minor projections ignored

Headroom stair

Minimum of 2 m. May be reduced for loft conversions in private dwellings.

Headroom landing

Minimum of 2 m. May be reduced for loft conversions in private dwellings.

Landings

Must be provided at the top and bottom of every flight. The length and width of landings must be at least the width of the stair; part of the floor may count as a landing. Landing must be free from permanent obstructions. A door may swing over the landing at the bottom of a flight, but must leave a clear space of 400 mm across the full width of the flight. Doors to cupboards and ducts may open in a similar way over landings at the top of a flight.

Guarding

Guarding in the form of a wall screen or balustrade is required to protect the sides of flights and landings in all buildings. Guarding need not be provided in private dwellings where the drop is 600 mm or less, or in other buildings where there are fewer than two risers. In buildings likely to be used by children under 5 years old, this guarding should not be easily climbed or permit the passage of a 100 mm sphere. Heights for all buildings are the same as for handrails, between 900 mm and 1000 mm measured vertically from the pitch line or floor to the top.

Handrail height and provision

Handrail required on one side of stair if under 1 m wide, and on both sides if over 1 m. Height for all handrails should be between 900 mm and 1000 mm measured vertically above the pitch line or landing to the top of the rail. Stairways in public buildings over 1800 mm wide should be subdivided by a handrail so that sub divisions do not exceed 1800 mm. Handrails are not required over the bottom two steps in private stairs, except when intended for disabled use.

Balustrading

Gap must not pass 100 mm sphere

Balustrading can be used to form the required guarding to the sides of flights and landings. Again in buildings likely to be used by children under 5 years old, the balustrade should not be easily climbed or permit the passage of a 100 mm sphere.

Open riser stairs

Down stand or rod to reduce gap — Minimum 16 mm overlap — Gap must not pass 100 mm sphere

Open risers permitted, but for safety the treads should overlap by at least 16 mm. In all stairways which are likely to be used by children under 5 years old the gap between treads must not permit the passage of 100mm sphere.

Tapered tread stairs

50 mm min. going — Less than 1 m — 50 mm min. going — 270 mm 270 mm — 1 m or more

The minimum going at any part of the tread within the width of the stair should not be less than 50 mm. The going is measured on the centre line of the stairs if they are not more than 1 m wide or at points 270 mm in from the ends of the treads for wider stairs. All consecutive treads should have the same taper. Where stairs contain both tapered and parallel treads, the going of the tapered treads should not be less than the parallel ones.

Stairs to loft conversions

Not less than 1800 mm — At least 1900 mm

Alternating treads may be permitted for loft conversions where there is no room for a proper staircase. They can only be used to give access to one room, must have handrails on both sides and non-slip treads.

Headroom may be reduced if the height at the centre of the stair is at least 1900 mm and not less than 1800 mm at the side.

Bore out at nosing end — Clean out nosing and saw edges of housing — Clean out waste

◀ **Figure 5.8** Sequence of operations for hand cutting

Figure 5.9 shows a method of forming a **bull-nose step**. The curved section of the riser is reduced to 2 mm thickness and bent around a laminated block. The wedges tighten the riser around the block and hold it there until the glue has set. The reduced section of the riser

▲ **Figure 5.9** Forming a bull-nose step

▲ **Figure 5.10** Splayed end step

▲ **Figure 5.11** Round end step (curtail)

should be steamed before bending. It can then be bent around the block fairly easily without risk of breaking. Splayed end steps are sometimes used in cheaper quality work with the riser mitred and tongued at the joints (see Figure 5.10).

A **curtail step** is illustrated in Figure 5.11. The shaped riser may be formed using the same method as used for bull-nose step, or alternatively it can be glulamed.

Figure 5.12 shows three alternatives for tread and riser details. The tread can be made from 25-mm timber and the riser traditionally from 19-mm timber or, as is now standard practice, 9-mm or 12-mm plywood is used for the risers.

Figure 5.13 shows how each step (tread and riser) is made up in a jig before being fixed to the strings. Glue blocks strengthen the joint between the tread and the riser. The absence or loosening of these often results in squeaky stairs.

A part view of the steps fixed into a string is illustrated in Figure 5.14. The treads and risers are glued and securely wedged into their positions in the string housing.

Figure 5.15 shows that the outer string and handrail are normally mortised into the newels at either end.

▲ **Figure 5.12** Tread and riser details

136

▲ **Figure 5.13** Step assembly jig

▲ **Figure 5.14** Fixing of steps into string

▲ **Figure 5.15** String handrail to newel joints

▲ **Figure 5.16**
Housing out newels

The newels will also require housing out to receive the treads and risers (see Figure 5.16).

Assembly of close string stairs

This on-edge assembly employs a cramping jig to cramp up the strings and hold the staircase square.

The following procedure is illustrated in Figure 5.17:

1. Clean up the inner housed faces of the string, the upper face and edge of tread and exposed face of the risers.

2. Pre-assemble the treads and risers including glue blocks.

3. Position one string on base of the cramping up frame.

4. Apply glue to the string housings and tread riser intersections of the pre-assembled steps. Place the steps in position (detail A).

5. Apply glue to the housings of the other string and place in position over the steps (detail B).

6. Close the cramping frame and tap up all the treads and risers, making sure they are all fully home in their housings (detail C).

7. Glue and drive the wedges for the treads. Trim off any surplus length with a chisel, so that they clear the riser housings.

A. Place pre-assembled step in string

B. Place other string over steps

Spreader beam

Angle frame

Folding wedges

C. Close jig and tap up steps in housings

Mecanically operated beam

Proprietary cramping jig

Angled frame jig bolted to bench

D. Wedge up, screw and brace underside

Piece of plywood fixed to underside to prevent raking

▲ **Figure 5.17** On-edge stair assembly

8. Glue and drive the wedges for the risers.
9. Screw bottom edge of risers to back edge of treads at about 225-mm centres.
10. Temporarily brace underside of stair, with diagonal braces or a piece of sheet material to prevent it racking out of square before installation (detail D).
11. Finally remove any squeezed out surplus glue from seen faces with a damp cloth. Clean up outer face and top edge of string. Remove any sharp arrises.

It is normal practice to assemble the flight of stairs to this stage only, for ease of handling and installation. Each flight will be separate, with the bottom bullnose step, top nosing, newels, handrail and balustrade supplied loose, ready for on-site completion.

Cut string stairs

Figure 5.18 shows the construction details of a cut and bracketed string. This type is used as the outer string of more decorative flights. The string is cut to the shape of the step and the treads are pocket screwed to them. The risers extend past the face of the string and are mitred with thin plywood brackets, which are glued and screwed in place.

Most of the assembly will be completed in the workshop with the exception of the following which will be completed on site: the balusters, which are dovetailed and screwed to the tread, and the return nosings, which are slot screwed to the end of the tread.

Open riser stairs

Shown in Figure 5.19 is a method used to join the treads to the strings in a riser-less flight of stairs (open plan). The through tenons should be wedged on the outside of the string. Alternatively, the treads could be simply housed into the string, and metal ties used under every third or fourth tread to tie the flight together.

Where carriages instead of strings are used in an open plan flight, the treads are supported by two timber brackets, which are either doweled or tenoned into the carriage as shown in Figure 5.20. Alternatively, these may be replaced by purpose-made metal brackets.

The balustrade of open plan flights is often of the straightforward **ranch style**. This is a number of planks either screwed and pelleted directly to the inside stair face of the newels or mortised and tenoned between them.

On certain flights where the gap between the treads is restricted a partial riser tongued to either the top or underside of the tread can be used. Figure 5.21 illustrates these two alternative methods.

▲ **Figure 5.18** Cut-and-bracketed stair

▲ **Figure 5.19** Open plan stairs

▶**Figure 5.20** Alternative bracket details

- metal bracket screwed to tread and carriage
- underside of tread housed to receive brackets
- bracket tenoned to carriage
- tread screwed and pelleted
- bracket dowelled to carriage

less than 100 mm

less than 100 mm

partial risers

▲ **Figure 5.21** Partial risers

Alternating tread stairs

A form of open riser stair having paddle-shaped treads, one part of the tread is cut away, with the wide portion alternating from one side to the other on each consecutive tread. They should only be used to give access to one room of a domestic **loft conversion**, where there is insufficient space to accommodate a proper staircase, see Figure 5.22. In addition a handrail should be fitted to both sides and the treads should have a non-slip surface

Tapered step stairs

Before marking out tapered steps they should be set out full size on a sheet of ply, as shown in Figure 5.23. This enables the shapes of the strings and tapered treads to be determined. The risers must normally radiate from a point outside the stair width in order to achieve the 50-mm minimum going. Shaped easing pieces are glued and tongued onto the string where the extra width is required. The two wall strings should be tongued and grooved together where they join.

The increased width of tapered steps necessitates them being made up from a number of pieces tongued together. Care should be taken when cutting to ensure that the nosing is kept parallel to the grain. Alternatively, the treads may be cut from sheet material with solid timber nosing glued to their front edge.

Geometric stairs

These have wreathed outer strings, which are normally cut and bracketed to receive the treads and risers. These strings consist of straight sections, joined either side of a curved section which rises and turns to suit the change in direction.

The two main methods of forming these wreathed strings are by glulam construction or staving. Staving involves reducing the curved section to about 2 mm so that it may be easily bent around a former or drum to

▲ **Figure 5.22** Alternating 'paddle tread' stairs

▲ **Figure 5.23** Setting out tapered steps

the required shape. After bending, timber staves are glued to the reduced section, thus building up the string to a uniform thickness and retaining it in its bent position.

Figure 5.24 shows a quarter turn **wreathed string** fixed around a drum ready for staving. After the staves have cured sufficiently the string may be removed from the drum, cut to the shape of the steps and prepared for jointing to the straight strings (see Figure 5.25)

These are jointed one rise past the turn with a cross-grained or loose tongue. A counter cramp can be used to pull the two ends tightly together as shown in Figure 5.26. This also has the effect of stiffening the stair and preventing movement at the joint. The cramp consists of three short pieces mortised to take the wedges. They are screwed to the strings initially only at one end. The two outside pieces are screwed to the wreathed string and the middle piece with its mortise slightly

▲ **Figure 5.24** Forming wreathed string

▲ **Figure 5.25** Wreathed string (staved and cut)

▶**Figure 5.26** Counter-cramp for joining geometrical stair strings

off-centre to the straight string. After driving the wedges to pull up the joint, the other ends can be screwed.

The handrail associated with a geometric stair also requires wreathing to follow the line of the string below. The setting out and construction of wreathed strings and handrails is a very highly skilled operation, carried out in general by only a few joinery works who specialise in purpose-made staircase manufacture. The geometrical setting out of such work is covered in Book 1.

Installation of stairs

The installation of any staircase on site follows the same basic procedure, although there will be differences depending on the type of stair and the nature of the site.

Stairs are normally delivered to site assembled as far as possible. However, for ease of handling each flight will be separate, its newels, handrail, balustrade and any tapered treads being supplied loose ready for on-site completion.

For maximum strength and rigidity the stairs should be fixed using the following details:

1. The top newel is notched over the landing or floor trimmer and either bolted or coach screwed to it (Figure 5.27).
2. The lower newel should be carried through the landing or floor and bolted to the joists.
3. Inserting a steel dowel partly into the newel and grouting this into the concrete can fix the lower newel on a solid ground floor (Figure 5.28).
4. The outer string and handrail are mortised into the newels at either end. With the flight in position these joints are glued and then closed up and fixed using hardwood draw pins.
5. The wall string is cut over the trimmer at the top and cut nailed or screwed to the wall from the underside see Figure 5.29.

▲ **Figure 5.27** Fixing outer string and handrail

▲ **Figure 5.28** Newel fixing to solid floors

◀ **Figure 5.29** Fixing wall string

6. The wall string is also cut at either end to provide an abutment for and suit the height of the skirting. All cut surfaces should be cleaned up with a block or smoothing plane, as there will be exposed end grain.

7. The bottom of a flight may be secured by screwing it to a batten fixed to the floor.

Finishing

Illustrated in Figure 5.30 is a section across a flight fixed up against one wall showing typical finishing details. Figure 5.31 shows how the trimmer around the stairwell opening is finished with an apron lining and nosing.

▲ **Figure 5.30** Section across flight open one side

▲ **Figure 5.31** Landing detail

Where the width of the stair exceeds about 1 m, a **carriage** may be fixed under the flight to support the centre of the treads and risers. To securely fix the carriage it is birdsmouthed at both ends, at the top over the trimmer and at the bottom over a plate fixed to the floor. Brackets are nailed to alternate sides of the carriage to provide further support across the width of the treads (see Figure 5.32).

▶**Figure 5.32** Carriage on wider stairs

Tapered treads

Any tapered treads are fixed after the main flights are in position. Figure 5.33 shows details of tapered steps at the bottom flight before and after the tapered treads have been completed. They should fit in place fairly easily as they will have been pre-fitted dry in the joiner's shop and disassembled for transportation. In practice a certain amount of adjustment is often required to take account of on-site conditions, e.g. slightly out-of-square or out-of-plumb brickwork etc.

◄ **Figure 5.33** Installation of tapered tread

After fitting the tapered treads and risers into the housings in the string, they should be wedged up and screwed. Because of their extra length some form of support is normally required under the tapered treads. This can be provided as shown in Figure 5.34 by bearers under the riser of each step. These are fixed between the two strings or between the string and the newel. Also shown is how the carriages of the main flights are birdsmouthed over these bearers.

Landings

The construction of landings and the materials used are normally similar to the upper floor. Figure 5.35 shows details of a half-space landing. The trimmer joist spans between two walls and supports the end of the trimmed joists spanning the landing.

Extending the newel post down to the floor, as shown in Figure 5.36, normally supports the free end of the trimmer for a quarter-space landing. Alternatively either a landing frame or some form of cantilever landing could be used.

Figure 5.37 shows a typical landing support framework. This would be subsequently boxed in or made into a cupboard, creating extra storage space.

The cantilever landing shown in Figure 5.38 consists of a diagonal joist built into the wall at both ends. This provides a bearing midway along the cantilever joist, which is built into the internal angle of the wall.

wall strings
tongued and
grooved
together at
corner

carriage to
upper flight
birdsmouthed
over bearer

bearers

bearers

carriage to
lower flight

▲ **Figure 5.34** Tapered treads

trimmer

trimmed

joist hanger

▲ **Figure 5.35** Half-space landing details

free end
of trimmer
supported
by newel

newel
post continued
down to floor

▲ **Figure 5.36** Quarter-space landing details

joists

support
frame

▲ **Figure 5.37** Landing support framework

The landing trimmer joists are dovetailed together and supported on their free end by the cantilevered joist. This method of support has the advantage of giving a clear space under the stair. It is essential that the timber used be well seasoned, as even slight shrinkage will cause the landing to move and creak in use.

Geometrical stair

These are supplied in a number of easily handled sections, the tapered treads being assembled on site as before. The main difference in the installation is the fixing of the 'wreathed' portion of the string. This will have been previously formed and permanently fixed at one end of the string. The other end of the wreathed string has to be fixed to its adjoining string on site.

The outer or wreathed string of a geometrical stair is usually of a cut-and-bracketed type, much of the assembly will have been completed prior to delivery with the exception of the balusters and return nosing. Where a section contains tapered treads they will normally be completely assembled on site.

Open-riser stair

The fixing of open-riser or open-plan stairs with close strings and newels is the same as the stairs previously mentioned. Where carriage or spine beam stairs are concerned the method of installation differs. Both types can be completely prefabricated in the joiner's shop, requiring only fixing at the top and bottom on site. Alternatively they can be delivered in knockdown form for on-site assembly.

Figure 5.39 illustrates a typical open-riser stair supported on carriages. The upper ends of the carriages are birdsmouthed around the trimmer, while the lower ends of the carriages are fixed to the floor with metal angle brackets. The treads are supported by and screwed and pelleted to timber brackets, which are in turn glued and dowelled to the carriages. Newel posts are fixed at either end of the flight onto which the ranch-style plank balustrading is screwed and pelleted.

Spine-beam stair

Illustrated in Figure 5.40 is a spine beam stair, which is also known as a mono-carriage stair. The spine beam is a large glulam section that is often tapered or curved on its underside to reduce its somewhat bulky appearance. Both ends of the spine beam are fixed to metal brackets that have been cast in the concrete; because of the great stresses and likelihood of movement at the junction between the tread and beam, it is essential that these be securely fixed. The shaped bearers are bolted, screwed or fixed with metal brackets to the beam. The treads can then be screwed and pelleted on to the bearers.

Metal rod balusters are often used for this type of stair. They are secured at one end to the tread. A mild steel core rail is used to secure the tops of the balusters. This is fixed using setscrews into the baluster

▲ **Figure 5.38** Cantilever landing

Labels on figure: wedges; cantilever joist wedged into internal corner; trimmer joists dovetailed; cantilevered diagonal joist; diagonal joist built in at both ends

▲ **Figure 5.39** Open-riser stair fixing details

plank balustrade screwed and pelleted to face of newel

newel bolted to trimmer

carriage bird's mouthed over trimmer

newel fixed to floor and bolted to carriage

treads screwed and pelleted to brackets

timber bracket pieces glued and dowelled to carriage

elevation

metal bracket fixing carriage to floor

section

▶**Figure 5.40** Spine beam stair

landing balusters set in concrete

hardwood handrail on mild steel core rail

mild steel balusters

metal bracket

treads on shaped bearers bolted to beam

glulam spine beam bolted at each end to metal brackets

metal bracket cast in concrete

elevations

after drilling and tapping. The hardwood handrail is grooved on the underside to conceal the core rail through which it is screwed. An enlarged detail of this is shown in Figure 5.41.

Treads of open-riser stairs are often found to be noisy and slippery in use. They can be made safer by fitting non-slip nosings to the front edges of the treads. Alternatively, to cut down on the noise and at the same time make them less slippery, carpet may be either wrapped around the treads between the carriages and tacked, or fitted into recesses that have been cut into the top of each tread (see Figure 5.42).

▲ **Figure 5.41** Balustrade detail

▲ **Figure 5.42** Safety features to open-riser stair treads

Balustrades and handrails

Balustrading

Stairs and landings can be fitted with balustrading in many ways other than the standard balusters, metal balusters and ranch-style plank balustrading covered in previous examples. Other methods of forming balustrades range from framed panelling with a variety of infill, including laminated or toughened glass, to decorative wrought ironwork in various designs.

Ranch-style plank balustrading is not recommended for buildings used by young people, as they can be easily climbed.

Standard balusters may be either stub tenoned into the string or fitted into a groove run into the string capping. At their upper end they are normally pinned into the groove run on the underside of the handrail (see Figure 5.43).

▲ **Figure 5.43** Fixing of balusters

plugged, screwed
and pelleted
to wall

screws
and pellets

plugged
and screwed
to wall

handrail screwed
through bracket

metal bracket
screwed to wall

▲ **Figure 5.44** Wall handrail
sections

Wall handrails

Handrails to straight flights with newels as illustrated in
previous examples are tenoned into the face of the newel posts.
In better-quality work they will also be housed into the face of
the newel by about 6 mm so that any shrinkage will not result
in an open, unsightly joint.

Wall handrails either may be fixed by plugging, screwing and
pelleting direct to the wall, or may stand clear of the wall on
metal brackets fixed at about 1-m centres. Figure 5.44
illustrates a traditional section and a modern built-up section
fixed direct to the wall, as well as one on brackets.

Where a handrail changes direction around a corner or from
rake to level, the section fixed directly to the wall may simply
be mitred. Handrails on brackets are not normally mitred but
change direction with the aid of short tangential curved
sections, jointed to the main straight lengths with handrail
bolts and dowels. The nuts are set from the underside of the
rail and the mortises are plugged with grain-matched inserts.
Figure 5.45 illustrates a number of these curved sections.

A quadrant is used to turn a level handrail around a 90° bend.
Ramps are either concave or convex and are used to join raked
to level handrails. Wreaths are double-curvature sections.
They are used where a raking handrail turns a corner, for
example in stairs with tapered steps, or where a raking
handrail turns a corner and changes to a level handrail, for
example at the junction of stair and landing. Alternatively a
ramp and mitre may be used. A half-newel can be used to give
support to the handrail and balustrade where it meets the wall
or a landing (see Figure 5.46).

The end of a handrail in a geometrical stair is supported by the
balusters and may either terminate in a scroll or be mitred into

level   joint   'quadrant'   level   rake   level   'ramp'

'wreath'   rake   level   level   mitre   'ramp'   rake

▶**Figure 5.45** Curved handrail
views

150

▲ **Figure 5.46** Return balustrades at landings

the top of a decorative turned newel post, as shown in Figure 5.47. The setting out and making of geometrical handrails is the work of a highly specialised joiner.

Where the balustrading is of metal, as is often the case in present-day non-domestic construction, the handrail is fitted to a metal core rail. It is common practice for the handrail to be wreathed, grooved and jointed and then given to the metalworker to produce a suitable core rail. Details of fixing metal balustrades and core rails have been illustrated in previous examples. Wall handrails may also be fixed to core rails for additional strength, see Figure 5.48.

▲ **Figure 5.47**
Geometric handrail
end details

▲ **Figure 5.48** Metal core rail and brackets

Timber facings

Concrete stairs can be given a more pleasing appearance by the addition of hardwood treads, as shown in Figure 5.49. These are screwed and pelleted to dovetail-shaped timber blocks that have been cast into the concrete.

Landings may be given the same finish by secret fixing hardwood tongued-and-grooved flooring to dovetail bearers that have again been cast into the concrete.

In situations where the riser has to be faced the detail shown in Figure 5.50 could also be used.

In certain circumstances a hardwood cut string is also used to finish the edge of a stair. This can be screwed and pelleted to cast-in fixing blocks.

hardwood treads and nosing screwed and pelleted to dovetail blocks cast in the concrete

hardwood T&G flooring secret fixed to dovetail bearers

dovetail block

▲ **Figure 5.49** Hardwood treads for concrete stairs

▲ **Figure 5.50** Hardwood risers for concrete stairs

Protection of completed work

hardboard or plywood pinned or taped in position

batten

▲ **Figure 5.51** Temporary protection of treads

After a new staircase has been installed, a short period spent taking measures to prevent damage during subsequent building work saves much more than it costs.

False treads made from strips of hardboard or plywood as shown in Figure 5.51 are pinned on to the top of each step. The batten fixed to the strips ensures the nosing is well protected.

On flights to be clear finished the false treads should be held in position with a strong adhesive tape, as pin-holes would not be acceptable.

Strips of hardboard or plywood are also used to protect newel posts. These can be either pinned or taped in position depending on the finish (see Figure 5.52).

Newel protected with
hardboard or plywood
pinned or taped in position

Handrails and balustrades
protected with corrugated
cardboard or bubble-wrap
held in place with tape

▲ **Figure 5.52** Temporary protection of newels and handrails

Adequate protection of handrails and balustrades can be achieved
by wrapping them in corrugated cardboard held in position with
adhesive tape.

Chapter Six

Doors, frames and linings

This chapter covers the work of both the site carpenter and bench joiner. It is concerned with principles of constructing and installing doors, frames and linings. It includes the following:

- an introduction to doors, frames and linings
- the performance requirements of doors
- methods of door construction and manufacture
- requirements of doorframes and linings
- hingeing and hanging of doors
- types of door ironmongery
- installation of door ironmongery.

Doors are moveable barriers used to cover an opening in a structure. Their main function is to allow access into and egress from a building, as well as the passage between the interior spaces. They may be classified by their method of construction and method of operation.

Introduction

Door frames and linings surround the wall opening on which doors are hung. The door frame is the surround on which an external door or internal door is hung consisting of two jambs, a head and sometimes a threshold and transom, normally with stuck-on solid stops and of a bigger section than door linings (Figure 6.1). The door lining is the surround on which mainly internal doors are hung, normally of a thinner section than door frames. Planted doorstops are normally used to accommodate the door, however stuck rebated stop linings are also available and termed as door casings. Architraves are fixed around door linings to cover the joint between the plaster and the wood. Door linings or casings are used for the majority of internal doors. See Figure 6.2.

The main difference between doorframes and door linings is that linings cover the full width of the reveal in which they are fixed from wall surface to wall surface whereas frames do not. Unlike doorframes,

Planted mould

Stuck mould

▲ Figure 6.1 Door frame

which are normally fully assembled in the workshop, it is common practice to produce linings in a knockdown form to be assembled on site by the carpenter.

►Figure 6.2 Door lining

Door performance requirements

The main function of a door is to allow access into and egress from a building, as well as the passage between the interior spaces. However, a door must be carefully designed and detailed so that it is capable of doing this while at the same time maintaining its other various performance requirements, such as weather protection, fire resistance, sound and thermal insulation, security, privacy, ease of operation and durability.

A door will rarely be expected to fulfil all of the performance requirements. In most situations doors are only expected to maintain a limited number of them. The priority given to each requirement will differ depending on the situation in hand.

- *Weather protection.* This requirement applies to external doors. These may be exposed to wind, rain, snow, sunlight and the extremes of temperature and must provide the same degree of weather protection as the remainder of the building. Openings are the weak point in a building as far as weather protection is concerned. Therefore careful consideration of materials to be used, construction details, hanging, glazing and ironmongery is critical.

- *Fire resistance.* Building elements that have a separating compartment or other fire protection role must be able to contain a possible fire and provide protection to the side remote from the fire for a given period of time. Doors that permit passage through this element must provide the same degree of protection in order to ensure a safe means of escape for the building's occupants. Doors in industrial buildings may also be required to protect the contents of the area until the fire has been extinguished.

- *Sound and thermal insulation.* In general, as the size of the door is relatively small in relation to the surrounding wall area, doors are rarely designed specifically for these requirements. Although the use of draught stripping, sealed unit double-glazing and solid rather than hollow construction can all be used to an advantage.

- *Security.* Door security depends on the materials used, soundness of construction, selection of suitable ironmongery and positioning. Only external door security is normally required for domestic dwellings, although internal door security is often necessary for offices, shops and factories, etc.

- *Privacy.* Unglazed doors provide total privacy when closed and if the layout has been carefully designed the door can still provide this privacy when partly opened. For this reason it is normal for doors in domestic dwellings to swing into the room rather than the wall. Glazed or partly glazed doors are used either to provide additional light through the doorway or to provide through vision for safety purposes. Partial privacy may be obtained by the use of obscure glass where through vision is not required.

- *Ease of operation.* A door's size, weight and position in a building will determine its method of operation, so correct selection of ironmongery is a most important design consideration.

■ *Durability*. A door, or any item of joinery for that matter, must be capable of giving satisfactory service, for a reasonable length of time, in the situation for which it was designed. In addition to the normal opening and closing, doors must also stand up to the occasional slamming and other misuse. Thus there is a need for sound, stable materials, rigid construction, suitable protection and again the selection and positioning of suitable ironmongery.

Door types

Doors may be classified by their method of construction: matchboarded, panelled, glazed, flush and fire resistant, etc. See Figure 6.3.

■ *Matchboarded doors* are used mainly externally for gates, sheds and industrial buildings. They are simply constructed from matchboarding, ledges and braces clench nailed together. The bottom end of the braces must always point towards the hanging edge of the door to provide the required support. Framed matchboarded doors constructed with the addition of stiles and rails are used where extra strength is required.

■ *Panelled doors* have a frame made from solid timber rails and stiles, which are jointed using either dowels or mortise and tenon joints. The frame is either grooved or rebated to receive one or more thin plywood or timber panels. Interior doors are thinner than exterior doors.

■ *Glazed doors* are used where more light is required. They are made similar to panelled doors except glass replaces one or more of the plywood or timber panels. Glazing bead is used to secure the glass into its glazing rebates. Glazing bars may be used to divide large glazed areas.

■ *Flush doors* are made with outer faces of plywood or hardboard. Internal doors are normally lightweight having a hollow core, solid timber edges and blocks, which are used to reinforce hinge and lock positions. New flush doors will have one edge marked 'LOCK' and the other 'HINGE'; these must be followed. External and fire resistant flush doors are much heavier, as normally they have a solid core of either timber strips or chipboard. A variation on flush doors is to use the same lightweight hollow core but have the faces covered in moulded or embossed facings to give the appearance of a traditional panel door. Internal doors use hardboard facings while plastic facings are mainly for external use.

◀ **Figure 6.3** Door types by construction method

157

Internal doors

Framed

Patt SA   Colonial   Patt
           pine   2XGG

Flush

White-faced   Plywood   Sapele
hardboard   (painted) (polished)

Sizes available
610 mm × 1981 mm × 35 mm
686 mm × 1981 mm × 35 mm
762 mm × 1981 mm × 35 mm
865 mm × 1981 mm × 35 mm

External doors

Patt 10   Patt 50   Patt SC   Patt 8

Carolina   Kentucky   Colonial   Half bow

Sizes available
762 mm × 1981 mm × 44 mm
835 mm × 1981 mm × 44 mm
813 mm × 2032 mm × 44 mm

▲ **Figure 6.4** Extract from a door manufacturer's door list showing the range of stock size doors

■ *Fire resisting doors* are mainly constructed as solid core flush doors. The main function of this type of door is to act as a barrier to a possible fire by providing the same degree of protection as the element in which it is located. They should prevent the passage of smoke, hot gases and flames for a specified period of time. This period of time will vary depending on the relevant statutory regulations and the location of the door. Fire doors are not normally purpose-made, as they must have approved fire resistance certification. It is advantageous to use proven proprietary products. Oversize fire door 'blanks' are available for cutting down to size, if required to suit specific situations.

Grading

The quality of a door will obviously affect its useful life (length of time that it is able to give satisfactory service). The majority of doors are therefore graded as being of either an **internal or external** quality.

Door sizes

All mass-produced doors may be purchased from a supplier in a range of standard sizes as shown in Figure 6.4. Special sizes or purpose-made designs are normally available to order from suppliers with joinery shop contacts.

In general internal doors have a finished thickness of either 35 mm or 40 mm. Whilst the thickness of external doors and fire doors are normally increased to 44 mm, in order to withstand the extra stresses and strains that they are subjected to.

While the type of adhesive used in manufacture is a significant factor to be considered when making purpose-made doors, in practice it is of less significance when using standard doors as most manufacturers use a synthetic adhesive for both grades. In addition, external doors should be preservative treated against fungal decay.

Matchboarded doors

This group of doors involves the simplest form of construction. They are suitable for use both internally and externally, although they are mainly used externally for gates, sheds and industrial buildings.

Ledged and matchboarded or ledged, braced and matchboarded doors

The basic type of door shown in Figure 6.5 consists of matchboarding, which is held together by ledges. This type is little used because it has a tendency to sag and distort on the side opposite the hinges. In order to overcome this, braces are usually incorporated in the construction (Figure 6.6). The use of braces greatly increases the rigidity of the door.

▲ **Figure 6.5** Ledged and matchboarded door

▲ **Figure 6.6** Ledged, braced and matchboarded door showing weathering if door is for external use

The bottom ends of the braces should always point towards the hinged edge of the door in order to provide the required support. Where these doors are used externally, the top edge of the ledges should be weathered to stop the accumulation of rainwater and moisture.

Three ledges are used to hold the matchboarding together. The outside pieces should be fixed with screws, while the remaining lengths of matchboarding are nailed to the ledges. Lost-head nails 6 mm longer than the thickness of the door are used for this purpose. The nails should be punched in and clenched over. Clenching over simply means bending the protruding part of the nails over and punching the ends below the surface (see Figure 6.7). The two braces, when used, are also fixed with lost head nails, which are clenched over. The joint detail between the ledges and braces is shown in Figure 6.8.

▲ **Figure 6.7** Clenching over

▲ **Figure 6.8** Joint details

Framed, ledged, braced and matchboarded doors

These are an improvement on the ledged, braced and matchboarded door, as they include stiles, which are jointed to the top, bottom and middle rails with mortise and tenons.

The use of the framework increases the door's strength, and resists any tendency to distort. Braces are optional when the door is framed, but their use further increases the door's strength.

Figure 6.9 shows the rear view of a typical framed, ledged, braced and matchboarded door. It can be seen from the section that the stiles and top rail are the same thickness, while the middle and bottom rails are thinner. This is so that the matchboarding can be tongued into the top rail over the face of the middle and bottom rails and run to the bottom

▲ **Figure 6.9** Framed, ledged, braced and matchboarded door   ▲ **Figure 6.10** Bare-faced tenons joint detail

of the door. As the middle and bottom rails are thinner than the stiles, barefaced tenons (tenons with only one shoulder) must be used. This joint is shown in Figure 6.10. These joints are normally wedged, although for extra strength draw pins can be used.

Non-standard matchboarded doors

Large **garage, warehouse and industrial doors** were often traditionally made using the framed, ledged, braced and matchboarded principles. This is because this type of door is ideal in situations where strength is more important than appearance. A pair of doors suitable for a garage is shown in Figure 6.11.

The portion of the door above the middle rail can be glazed to admit a certain amount of light into the building. The top, middle and glazing rails must be rebated in order to receive the panes of glass which are

▶ **Figure 6.11** Part-glazed garage doors

held in the rebate by glazing sprigs and putty. When the upper portion of the door is glazed the middle rail will be the same thickness as the stiles and therefore bare-faced tenons are only used on the bottom rail.

The joint normally used at the intersections of the glazing rails is a scribed or mitred cross-halving see Figure 6.12. As an alternative a stub tenon can be used, but as the glazing rails are of a small section the joint is not as strong as the halving.

A section through the meeting stiles of a pair of garage doors is illustrated in Figure 6.13, along with an alternative means of joining the matchboarding to the stiles.

Very large industrial doors are normally made to slide rather than be side hung, because of their increased tendency to sag as weight and size are increased.

Details of a pair of large industrial doors are illustrated in Figure 6.14. A small side-hung wicket door has been included to enable personal access without the need to open the main doors, although this is not common to all such doors. The main difference in the construction of these large doors is the increased sectional size of the members. In addition, since they are designed to slide, they are cross-braced to resist the tendency to distort when sliding in either direction.

A pair of external **Gothic design doors** made in oak and suitable for church use is illustrated in Figure 6.15. The internal and external elevations of each leaf are indicated in the one illustration. The construction of these doors differs only in the shaped head from the standard framed, ledged and braced details.

▲ **Figure 6.12** Glazing rail or bar: joint detail

▲ **Figure 6.13** Rebated meeting stiles and matchboarding to stile detail

▲ **Figure 6.14** Industrial sliding doors

70 mm × 145 mm top rail
70 mm × 145 mm stile
45 mm × 145 mm brace
25 mm matchboarding
45 mm × 220 mm middle rail
45 mm × 220 mm bottom rail

inside
side-hung wicket door
outside

false
tenon

bridle
joint

outside   inside

▶ **Figure 6.15** Gothic head doors

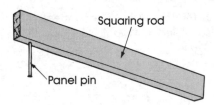

Squaring rod

Panel pin

▲ **Figure 6.16** Squaring rod

Squaring rod used to
check both diagonals

▲ **Figure 6.17** Checking the frame
for square

A pinned bridle joint is used between the outside stile and top rail.
A false tenon is inserted to join the top rail to the meeting stile, as the
rail's short grain at this point discounts the use of a normal tenon.
In addition, the false tenon considerably strengthens the rail's short
grain and minimises the possibility of it shearing.

Assembly procedure for framed matchboarded doors

The main framework should first be assembled dry to check the fit of
joints, sizes, square and winding.

Squaring up of a frame

This is checked with a squaring rod, which consists of a length of
rectangular section timber with a panel pin in its end as shown in
Figure 6.16. The end with a panel pin is placed in one corner of the
frame (see Figure 6.17). The length of the diagonal should then be
marked in pencil on the rod. The other diagonal should then be
checked. If the pencil marks occur in the same place, the frame must be
square. If the frame is not square, then sash cramps should be angled to
pull the frame into square as shown in Figure 6.18.

Winding of a frame

This is checked with winding strips (see Figure 6.19). These are two
parallel pieces of timber. With the frame laying flat on a level bench,
place a winding strip at either end of the job. Close one eye and sight
the tops of the two strips as shown in Figure 6.20. If they appear parallel
the frame is flat or 'out of wind'. The frame is said to be winding, in
wind or distorted if the two strips do not line up. Repositioning of the
cramps or adjustment to the joints may be required.

◄ Figure 6.18
Pulling the frame into
square using cramps

▲ Figure 6.19 Winding strips

▲ Figure 6.20 Winding strips used to check frame for winding

Glue up, assemble and lightly drive wedges

A waterproof adhesive should be used for external joinery or where it is likely to be used in a damp location. Ensure the overall sizes are within the stated tolerances. Re-check for square and wind. Assuming all is correct, finally drive in the wedges.

Figure 6.21 shows how the stiles and rails are assembled, glued and wedged before the matchboarding is fixed.

Before the assembly of any matchboarded door, all concealed surfaces, such as the tongues and grooves of the matchboarding, back of rails, ledges and braces, should be coated either with a priming paint for painted work, or with a suitable sealer where a clear finish is required. Alternatively preservative-treated timber should be used. This is in order to prevent moisture penetration and subsequent decay of the timber.

- Arrange the boards so that the two outside ones are of equal width. They may either be tongued into the top rail and stiles or simply fit into a housing (see Figure 6.22).
- Lay the boards on the assembled frame. Locate tongues and grooves so that the boards form an arc between the jambs. Place a short piece of timber across the door at either end. With assistance apply pressure at both ends of the door to fold the boards flat (see Figure 6.23).

Wedges to be driven in order indicated to ensure correct positioning of rails

Sash cramp either side of middle rail

Level bench bearers

Timber packing behind cramp head to avoid damage

Cramps kept clear of mortise to enable wedging

Apply waterproof glue to exposed faces of tenon and shoulders

Level bench bearer

Rails and stiles loosely assembled with tenons only just inserted

▶ **Figure 6.21** Assembling framed matchboard door

Outside boards of equal width

Rail   Stile

Boards either housed or tongued into stiles/top rail   Matchboard

▲ **Figure 6.22** Matchboarding details

Apply hand pressure to fold boards flat

Tongues and grooves located to form an arch

▲ **Figure 6.23** Folding matchboards

Brace cut in corner of framing   Brace let into rail

▲ **Figure 6.24** Fitting braces

- Tap up the boards from the bottom to locate them correctly into the top rail and then clench nail or staple them (using a pneumatic nail gun) to all framing members.
- Mark, cut and fix the braces. These may be either cut into the corners of the framework, or let into the rails (see Figure 6.24). The cut in the corner method is simpler; however, it has a tendency to push open the joints between the stiles and the rails.

Panelled and glazed doors

The design and construction of panelled doors are very similar to those of glazed doors. They consist of a frame which has either a plough groove or rebate run around it to receive the panels or glazing. The framing members for these doors vary with the number and arrangements of the panels. They will consist of horizontal members and vertical members.

Rails, stiles and muntins

All horizontal members are called rails. They are also named according to their position in the door, such as top rail, middle rail, bottom rail, intermediate rail. The middle rail is also known as the lock rail and the upper intermediate rail is sometimes called a frieze rail. The two outside vertical members are called stiles, while all intermediate vertical members are known as muntins.

A typical panelled door, with all its component parts named is shown in Figure 6.25. The middle and bottom rails are of a deeper section as they serve to hold the door square and thus prevent sagging. Muntins are introduced in order to reduce the panel width, therefore reducing the unsightly effect of moisture movement and the likelihood of panel damage.

It is normal to leave at least a 50-mm horn on each end of the stiles. This serves two purposes:

- It enables the joints to be securely wedged without fear of splitting out.
- The horns protect the top and bottom edges of the door before it is hung.

Panels and mouldings

Figure 6.26 shows a ply panel which is held in a plough groove that is run around the inside edge of the framing. Two **ovolo mouldings** are also worked around the inside edges of the framing for decorative purposes. They are known as **stuck mouldings**. The plough groove should be at least 2 mm deeper than the panel. This is to allow for any moisture movement (shrinkage and expansion).

Figure 6.27 again shows a solid or plywood panel which is held in a plough groove that is run around the inside edge of the framing. Here a **planted mould** has been applied around the panel for decoration. This method avoids the need to scribe or mitre the shoulders of the rails, which applies with stuck mouldings. Planted moulds must not be allowed to restrict panel movement. Therefore they should be pinned to the framing and not the panel.

Figure 6.28 shows a timber panel which is tongued into a plough groove in the framing. This type of panel is known as a **bead butt panel** because on its vertical edges a bead moulding is worked, while the horizontal edges remain square and butt up to the rails.

Figure 6.29 shows a glazed panel which is located in the rebate. It is held in position by **planted beads**, which are pinned into the framing. Where this type of door is used externally the planted beads should be replaced by face putty or placed on the inside of the door mainly for security reasons but also because when glazing beads are used externally water tends to get behind them. This makes both the beads and framing susceptible to decay. The procedure for glazing using face putty and beads is covered in chapter 12 of this book.

45 mm × 95 mm top rail

45 mm × 95 mm muntin

45 mm × 95 mm frieze rail or intermediate rail

45 mm × 95 mm muntin

30 mm panel

45 mm × 195 mm middle or lock rail

45 mm × 95 mm muntin

45 mm × 95 mm stile

45 mm × 195 mm bottom rail

50 mm horn

▲ **Figure 6.25** A typical panelled door

Stuck ovolo moulding

Ply or solid panel

▲ **Figure 6.26** Panel detail (stuck moulding)

Planted mould

Ply or solid pane

▲ **Figure 6.27** Panel detail (planted mould)

Bead and butt panel

▲ **Figure 6.28** Panel detail (bead butt)

Ply or glazed panel

Pinned glazing bead

▲ **Figure 6.29** Panel detail (planted bead)

▲ **Figure 6.30** Panel detail (bolection mould)

Figure 6.30 shows a planted mould that is rebated over the framing in order to create an enhanced feature. This type is known as a **bolection mould**. In general bolection moulds are fitted on the face and planted bead moulds on the reverse, although in the case of top quality work bolection moulds could be used on both faces.

The bolection mould is fixed through the panel with screws. The holes for the screws should be slotted across the grain to permit panel movement without risk of splitting. The planted bead mould used on the other side to cover the screws should be skew nailed to the framing.

In good quality joinery, refurbishment or restoration work, the panels themselves may be decorated by working various mouldings on one or both of their faces. The portion around the edge of a panel is called the margin and the centre portion is known as the field. The small flat section around the edge of the panel is to enable its correct location in the framing.

Decorative panels

Figure 6.31. Illustrates the section and part elevation of the main types of decorative panels:

(a) *Raised or bevel raised panel.*

(b) *Raised and fielded panel,* also known as bevel raised and fielded, where the margin has been bevelled to raise the field.

(c) *Raised, sunk and fielded panel,* also known as bevelled, raised sunk and fielded. In this case the margin has been sunk below the field to emphasise it.

(d) *Raised, sunk and raised fielded panel,* also known as bevelled raised, sunk and bevelled raised fielded panel, where the field itself has been bevelled as a further enhanced detail.

When bolection moulds or planted moulds are used to finish decorated panels, the small flat section around the panel edge must be extended to provide a flat surface that will accommodate the mouldings.

▲ **Figure 6.31** Decorative panels: (a) raised; (b) raised and fielded; (c) raised, sunk and fielded; (d) raised, sunk and raised fielded

Joints

Traditionally the mortise and tenon joint was used exclusively in the jointing of panelled and glazed doors, but today the majority of doors are mass produced and in order to reduce costs the dowelled joint is used extensively.

Dowelled joints

The use of the dowelled joint reduces the cost of the door in three ways:

- The length of each rail is reduced by at least 200 mm.
- The jointing time is reduced, as holes only have to be drilled to accommodate the dowel.
- The assembly time is reduced, as no wedging, etc., has to be carried out.

Figure 6.32 shows a six-panel door, which has been jointed using dowels. These dowels should be 16 mm × 150 mm and spaced approximately 50 mm centre to centre. The following is the minimum recommended number of dowels to be used for each joint:

- Top rail to stile: two dowels
- Middle rail to stile: three dowels
- Bottom rail to stile: three dowels
- Intermediate rail to stile: one dowel
- Muntin to rail: two dowels

Figure 6.33 shows an exploded view of a doweled joint between a top rail and stile. In addition to the dowel, a haunch is incorporated into the joint. This ensures that the two members finish flush. Its use also overcomes any tendency for the rail to twist.

▲ **Figure 6.32** Dowel-jointed panel door

◄ **Figure 6.33** Exploded view of dowelled joint

Dowels should be cut to length, chamfered off at either end to aid location and finally a small groove is formed along their length, to allow any excess glue and trapped air escape when the joint is cramped up. Alternatively 'ready-made' dowels may be used. These are available in a range of sizes; they have chamfered ends and multi-grooved sides.

Mortise and tenon joints

Although the dowel joint is extensively used for mass-produced doors, the mortise and tenon joint is still used widely for purpose-made and high-quality door construction.

Figure 6.34 shows an exploded view of the framework for a typical six-panel door. Haunched mortise and tenons are used for the joints between the rails and stiles.

For joints between the muntins and rails, stub mortise and tenons are used. As these joints do not go right through the rails, they cannot be wedged in the normal way. Instead fox wedges are used (see Figure 6.35). These are small wedges, which are inserted into the saw cuts in the tenon. When the joint is cramped up the wedge expands the tenon and causes it to grip securely in the mortise.

▲ **Figure 6.34** Exploded view of mortise and tenon joints

▲ **Figure 6.35** Fox wedge joint

Half-glazed door

A traditional half-glazed door is shown in Figure 6.36. It is constructed with diminishing stiles, in order to provide the maximum area of glass and therefore admit into the building the maximum amount of daylight. This type of door is also known as gun stock stile door because its stiles are said to resemble the stock of a gun. The middle rail has splayed shoulders to overcome the change in width of the stiles, above and below the middle rail. An exploded view of this joint is shown in Figure 6.37.

▲ **Figure 6.36** Half-glazed door with diminishing stiles

Labels (top to bottom): Top rail · Glazing bars · Diminished stile · Middle rail · Panel with bolection mould · Bottom rail

▲ **Figure 6.37** Diminished stile joint detail

Labels: Moulding mitred · Splayed shoulder

The top half of the door can either be fully glazed or subdivided with glazing bars as shown. When glazing bars are used they are normally stub tenoned into the stiles and rails. The joints between the glazing bars themselves are the same as used for the upper glazed portion of the matchboarded door, either stub tenoned or halved and scribed.

The bottom half of the door normally consists of a bevel raised sunk and fielded panel with planted bolection mouldings.

Assembly procedure for panelled and glazed doors

This can be carried out for all types of framed door, using the following assembly procedure:

1. Dry assemble to check fit of joints, overall sizes, square and winding.
2. Clean up inside edges of all framing components and both faces of panels.
3. Glue, assemble, cramp up and wedge. Re-check for square and winding.
4. Clean up remainder of item and prepare for finishing.

The assembly of a six-panel door is illustrated in Figure 6.38. The rails and muntins should be glued and assembled first. Panels can then be inserted dry, taking care to ensure that no glue has squeezed into the panel grooves. Next the stiles are positioned, glued, cramped and wedged followed by final cleaning up.

169

Assemble rails and muntins

Bench bearer

Use packing
behind cramp
to avoid damage

Insert panels

Ensure panel
groove is free
of glue

Wedges to be driven
in order indicated to
ensure correct
positioning of rails

Position stiles
glue, cramp
and wedge up

Top
rail

Intermediate
rail

Middle
rail

Stile

Bottom
rail

Apply
glue to
both faces
of tenon
and shoulders

Rail

Stile

Rail just inserted for glueing up

▲ **Figure 6.38** Assembling six-panelled door

Flush doors

Joinery works are rarely involved with the manufacture of flush doors, except where special features, not available in standard mass-produced doors, are required.

Flush doors consist of either a hollow or solid core that is faced with sheets of hardboard, plywood or moulded plastic.

Hollow core doors

These are normally the cheapest type of flush door to produce and purchase. There are many ways in which the hollow core can be made. The two main methods are the skeleton core or the lattice core.

Skeleton core door

Figure 6.39 shows a skeleton core door, which is suitable in one-off or limited production. It consists of 28 mm × 70 mm stiles, top and bottom rails; 20 mm × 28 mm intermediate rails are used to complete the framework.

There is a tendency for the facing to deflect over time between the compartments of the skeleton core door. This produces a ripple effect, which is especially noticeable when the facings are finished with a gloss coating such as paint or varnish.

▲ **Figure 6.39** Skeleton core door construction

171

▲ Figure 6.40 Lattice core door construction

Very simple joints can be used in this type of construction, as its main strength is obtained by firmly gluing the facings both to the framework and to the core. The rails are usually either tongued into a groove in the stiles or butt jointed and fixed with staples or corrugated fasteners. Ventilation holes or grooves must be incorporated between each compartment. This is to prevent air becoming trapped in the compartment when the door is assembled. If this were not done the facings would have a tendency to bulge.

Lattice core door

A lattice core door is shown in Figure 6.40. It consists of 28 mm × 30 mm framework, which is simply stapled together. The core is made from narrow strips of hardboard slot jointed to produce the lattice. As with other types of flush door construction the strength of the door is obtained by firmly gluing the facings to both the framework and the core.

Honeycomb core door

Moulded panel doors which are currently very popular, use a hollow core door construction with the application of facings that are pressed or moulded to give the impression of a traditional panelled door See Figure 6.41. The core shown is typical of those that are used in mass production. It is formed from cardboard in a honeycomb pattern. The facings are moulded from hardboard for internal doors, and are normally impressed with a textured wood grain pattern, which is primed ready to receive a paint or wood stain finish. Doors with moulded plastic facings are available for external use.

A **lock block** should be provided in all hollow core doors to accommodate a mortise lock or latch. The block also serves to provide a fixing for the lock furniture. The position of the lock block is normally indicated on the edge of the door.

Solid core door

Solid core flush doors are considered to be of a better quality than hollow flush doors. There are four main reasons for this view:

- their facing remains flat;
- their increased rigidity;
- their increased levels of sound and thermal insulation;
- their increased resistance to penetration by fire.

Laminated core door. Figure 6.42 shows a laminated core door. It consists of 25-mm strips, which are firmly glued together along with the two plywood facings. The strips should be laid alternately in order to balance any stresses.

Chipboard core door. Figure 6.43 shows a chipboard core door. It consists of the chipboard core, which is surrounded by a simple 30-mm framework. The frame-work is normally stapled together at the corners. Gluing the facings to either side completes the door. Extruded tubular core chipboard could be use as an alternative, where a lighter weight door is required.

Simple softwood framework

Honeycomb pattern cardboard core

Moulded hardboard facing, which is normally impressed with a woodgrain pattern

▲ Figure 6.41 Moulded panel door with honeycomb core

Simple softwood
frame, often stapled
at corners

Solid
chipboard
core

Lipping

Extruded tubular
core chipboard
may be used for
a lighter door

Plywood or
hardboard facing

▲ **Figure 6.42** Laminated core door

▲ **Figure 6.43** Chipboard core door

Lippings

These are narrow strips of timber, which are fixed along the edges of
better quality flush doors. Their purpose is to mask the edges of the
facings and provide a neat finish to the door. External doors should
have lippings fixed to all four edges for increased weather
protection.

Plain lipping is acceptable for internal doors, but for better quality
internal doors and external doors tongued lipping is preferred, see
Figure 6.44. Lipping should be glued in position and not fixed with
panel pins.

▲ **Figure 6.44** Flush lipping and
tongued lipping

Vision panels

These are often required in purpose-made flush doors. They provide a
view of the other side, reducing the risk of collisions when the door is
opened, as well as the transmission of light. Additional framing is
required around the opening for hollow core doors as shown in
Figure 6.45. Shaped blocking out pieces can be used to form circular
vision panels (see Figure 6.46). This framing and blocking is normally
done during the construction of the door, although in the case of
simple rectangles, it may be possible to form the opening at a later
stage. Vision panels in fire doors may be restricted in both size and
shape, in order to comply with its certification.

Glazing beads

These are used to hold the glass in place. Plain glazing beads are
suitable for internal doors, but for increased protection from the
elements on external doors they should be rebated (bolection) and

▲ **Figure 6.45** Flush door with vision panel

◀ **Figure 6.46** Blocking for circular vision panel

Shaped blocking

▲ **Figure 6.47** Internal and external glazing beads

cupped and screw glazing bead

internal face

external face

▲ **Figure 6.48** Alternative external flush door glazing detail

the top surface weathered (Figure 6.47). A better method for fixing external glazing beads is illustrated in Figure 6.48. As well as providing the increased weather protection of the previous method, it also provides far greater security as the beads can't be removed from the outside.

Glazing beads for doors are preferably fixed using countersunk brass screws and recessed cups on at least one face rather than nails or pins. This is to enable the glazing beads to be easily removed in the event of the glass needing replacement.

Safety glazing

Safety glass should be used to glaze the critical areas of doors and other vulnerable low areas, particularly where children are present The Building Regulations state that glass in these areas should either: break safely (does not produce pointed shards with razor sharp edges); be robust (adequately thick to prevent the likelihood of breaking); or be permanently protected (by screens).

▲ **Figure 6.49** Glazing to doors, windows and screens

The critical areas and Building Regulations requirements are illustrated in Figure 6.49. Toughened and laminated glass can meet the requirements. Where annealed glass is used it must be at least 6 mm thick and in panes not larger than 0.5 m² with a maximum width of 250 mm. Screens or railings on both sides, which are at least 800 mm high, should be used to permanently protect fixed annealed glass.

Other types of door

Double margin door

This is a pair of narrow doors joined together at their meeting stiles to make a single door, with the appearance of a pair (see Figure 6.50). It is used for very wide openings where a single door would give an ill-proportioned, awkward appearance and where each half of a double door would not allow easy pedestrian passage.

The correct procedure for assembly is to glue and wedge the rails to the meeting stiles and then join together the two stiles with folding wedges. The panels are inserted from either side before assembling the outer stiles, cramping up and wedging. Figure 6.51 shows an exploded view of the framework for a typical double margin door.

For additional stiffness, a flat metal bar can be housed and screwed across the joint at both the top and bottom edges. Traditionally the top and bottom rails would be continuous across the whole door width, with the meeting stiles bridle jointed to them.

▲ **Figure 6.50** Double margin door

▲ **Figure 6.51** Exploded view of double margin door

Figure 6.52 shows a section through the meeting stiles. It illustrates the positioning of the folding wedges. In addition, it shows that the stiles are rebated together and a stuck bead is used to break the joint between the stiles. Alternatively the two stiles can be tongued together with a loose tongue and the joint masked with either stuck or planted beads.

Double-faced doors

In certain situations, particularly in prestigious public buildings and stately homes, etc., adjacent rooms or corridors may be panelled in a different timber or to a different detail. In these cases it is necessary for the face of the door to match in timber and design the room or corridor in which it is situated. With flush veneered panelling all that is required is the application of different veneer. Where a framed panel door is required the solution is to use two thin doors, each about 28 mm thick, which are fitted to each other with dovetail keys to form a door that matches its respective location (see Figure 6.53).

▲ **Figure 6.52** Meeting stile detail

The dovetail keys are screwed at intervals to the backs of the framing of one door and housed into the framing of the other (see Figure 6.54). The corresponding members (stiles, rails etc.) of each door face are paired, glued and keyed together before the door itself is assembled.

After assembly the edges of the door must be finished with a veneer or mitred lipping. The closing stile lipping should be matched to the room in which it opens, while the hanging stile should be matched to the other.

▲ **Figure 6.53** Detailed key for double-faced doors

▲ **Figure 6.54** Double-faced door finishing detail

Curved rail doors

The main problem encountered with the construction of curved work is that of short graining. This causes a weakness in the member because of a marked tendency to sheer along the short grain. This risk can be reduced by taking care when marking out the member to ensure that there is as little short grain as possible (see Figure 6.55).

The design of the joint is also important, as this is often the area most affected. Figure 6.56 illustrates the joint between a shaped top rail and stile. It shows how the shoulder is diminished in order to thicken up the short grain section of the rail.

The elevation of a semicircular-headed door is shown in Figure 6.57. The door head is formed in two parts. Traditionally hammer-headed joints would have been used, see Figure 6.58. A double-ended hardwood hammer-headed key is used at the crown. The wedges secure the joint and draw the two parts together. A hammer-headed tenon is used to join the stile and curved head. This is the same as one end of the key except that it is formed on one end of the stile.

An alternative method of jointing which reduces the amount of handwork is illustrated in Figure 6.59. This shows how the crown can be jointed using a false tenon insert and a bridle joint between the stile and rail. Both joints are secured using hardwood draw pins.

▲ **Figure 6.55** Marking out curved members

▲ **Figure 6.56** Shaped rail joint detail

▲ **Figure 6.57** Semicircular-headed door

▲ **Figure 6.58** Traditional joint details, semicircular-headed door

▲ **Figure 6.59** Alternative joint details, semicircular-headed door

Whichever method is used to construct the semicircular-headed door, the immediate rail should be kept about 50 mm below the head/stile joint. This is in order to ensure sufficient strength and minimise the possibility of shearing along the short grain between the mortise and shoulder.

Louvred doors

The frame of this type of door is similar in construction to that of a panel door, that is, mortise and tenons or dowels are used to join the rails to the stiles. Traditionally, louvres were only used as a means of ventilation, but now their use is increasing because of their pleasing appearance.

External louvred doors. When used externally, the louvre slats or blades are normally set at an angle of 45°. They should also project beyond the face of the framing for weathering purposes. Figure 6.60 shows an elevation and section of an external half-louvred door.

There are two methods by which the projecting louvre slats can be jointed to the stiles. Figure 6.61 shows the first method where the stiles have been mortised to receive the tenons, which have been formed on the ends of the louvred slats. Figure 6.62 shows the second method. Here the stiles have been through-housed to accommodate the full thickness of the louvre slats.

In all cases the slats should have an overlap of 6 mm to prevent through vision.

Internal louvred doors. Where louvre doors are used internally it is not necessary for the louvre slats to project beyond the face of the framing. The internal slats are normally set at an angle of 60° for two main reasons: more of the face of the slat is visible, adding to the decorative appearance, and fewer slats are required, giving a more economical door.

Figure 6.63 shows the method normally used for jointing the louvre slats of an internal door to the stiles. The stiles are stop-housed on a high-speed router using a purpose-made jig. The housing must be cut to exactly the same size as the louvre slat, otherwise the slat will be a slack fit, or time will be wasted easing the housing to fit each individual slat.

▲ **Figure 6.60** Louvre door

▲ **Figure 6.61** Louvre slat joint detail (mortise and tenon)

▲ **Figure 6.62** Louvre slat joint detail (though housing)

▲ **Figure 6.63** Louvre slat joint detail (stopped housing)

Fire resisting doors

As stated previously this type of door is not normally purpose made, as they must have approved fire resistance certification. This entails submitting the door assembly, pre-hung in its frame and fitted with all its intended ironmongery, to an approved fire-testing organisation.

They will test the fire resisting door assembly by reference to the following criteria:

- *Stability:* resistance to the collapse of the door.
- *Integrity:* resistance to the passage of flames or hot gases to the unexposed face.
- *Insulation:* resistance to the excessive rise in temperature of the unexposed face.

All fire-resisting doors are now called 'fire doors' (FD). Traditionally they were prefixed by their stability and integrity rating respectively. Thus a 60/45 fire-resisting door had a minimum 60 minutes stability rating and a 45 minutes integrity rating.

The more commonly used terms for fire doors were 'fire check' and 'fire resisting'. 'Fire check' was used to signify doors with a reduced integrity. For example 60/60 and 30/30 were known as 1 hr and $\frac{1}{2}$ hr fire-resisting doors; 60/45 and 30/20 were known as 1 hr and $\frac{1}{2}$ hr fire check doors.

Today, all fire doors are termed by their integrity rating only, e.g. FD20, FD30 and FD60. A suffix 'S' may be added to a specification, to indicate a fire door that requires smoke seals fitted for example FD30S.

Fire doors should be identified by a permanent label or a colour coded plug inserted into their hanging edge, see Figure 6.64. White for FD20; yellow for FD30 and blue for FD60. Plugs are further identified with a coloured central core or spot. Red indicates that an intumescent seal must be fitted on site; green indicates that no additional intumescent seal is required. A blue core may also be used on a white background for FD20 doors without intumescent seals and FD30 doors with an intumescent seal.

The actual fire door required will vary depending on the relevant statutory regulations and the location of the door. For instance, it is significant if the door is providing access through a separating or compartment wall, opening on to a protective shaft, or separating a hazard (attached garage, boiler house, etc.). Therefore, consultation with the Local Authority's Fire Prevention Officer and the Building Regulations is advised in order to determine the specific requirements at the design stage.

Fire door seals

The weak point in fire door construction is the joint between the door and frame. This is where the fire and smoke will penetrate first. Intumescent strips or seals are required for many fire doors. They may

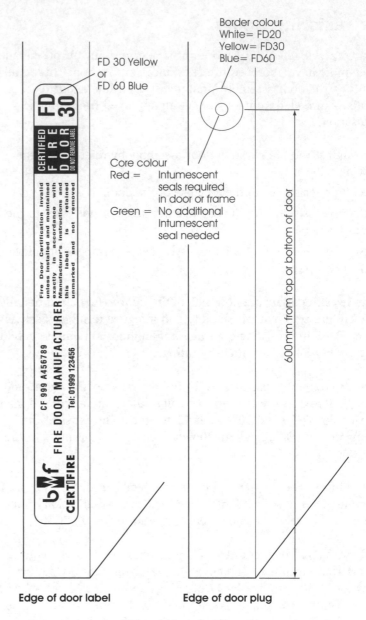

FD 30 Yellow
or
FD 60 Blue

CERTIFIED
FIRE
DOOR
DO NOT REMOVE LABEL

FD
30

Fire Door Certification invalid
unless installed and maintained
exactly in accordance with
Manufacturer's instructions and
this label is retained
unmarked and not removed

CF 999 A456789
FIRE DOOR MANUFACTURER
Tel: 01999 123456

bwf
CERTIFIRE

Border colour
White= FD20
Yellow= FD30
Blue= FD60

Core colour
Red =   Intumescent
         seals required
         in door or frame
Green = No additional
         Intumescent
         seal needed

600mm from top or bottom of door

Edge of door label   Edge of door plug

▲ **Figure 6.64** Identification of fire doors

Frame
Gap

Door

Gap
Seal in normal use

Gap sealed

Door

Gap sealed
Seal expanded after heat activation

Intumescent
seal/strip in
door frame jambs
and head

▲ **Figure 6.65** Fire door seals

be fitted into the door lipping, around the doorframe or to both the door and the frame; however this must always be in accordance with its test certification. When activated by high temperatures in the early stages of a fire these strips expand, sealing the joint and prolonging the door's integrity see Figure 6.65.

Expanded strips also serve to lock the door into its frame, thus prolonging its stability. In the early stages of a fire before the activation of the intumescent strips, the inhalation of smoke and fumes passing through the gap around the door presents the major hazard, particularly on designated escape routes. These doors should be fitted with smoke seals as a preventative measure. Figure 6.66 shows the two main types of seal: compression seals are normally fitted into the doorstop, and wiping seals can be fitted into either the door edge or the frame itself.

Vision panels

These are permitted in fire doors provided they have been approved and tested. In fact vision panels in fire doors have several advantages. In addition to their general use, where they enable a person to see if the passage is clear on the other side, in the event of a fire they provide a visual warning before the door is opened. Georgian wired or a special fire-resisting safety glass is normally specified. This should be held in place using the specified method, untreated timber beads with the glass bedded in intumescent paste is normally suitable for lower rated doors, although coating the beads with intumescent paint, or capping with a metal angle gives greater protection. Non-combustible glazing systems are recommended for higher rated doors, as timber beads may not give the required levels of integrity. Figure 6.67 illustrates typical fire door glazing details. However it must be remembered that any glazing to a fire door, can only be undertaken if it conforms to its fire test certificate. Full details should be available from the door manufacturer.

▲ **Figure 6.66** Smoke control door seals

◄ **Figure 6.67** Typical fire door vision panel glazing details

Methods of door operation

A door's method of operation will be determined by its location, construction and desired performance requirements. Figure 6.68 illustrates various methods of door operation:

- swinging
- sliding
- folding
- revolving.

swinging, side hung

single-leaf   single swing

single-leaf   double swing

double-leaf   single swing

double leaf   double swing

sliding

straight run   single track

straight run
bipass

curved run   single track

up-and-over

folding

end folding

folded
position

centre folding

revolving

▲ **Figure 6.68** Methods of door operation

Swinging doors

These are the most suitable doors for pedestrian use and the most effective for weather protection, fire resistance, sound and thermal insulation. Side hanging on hinges is the most common means of door operation, although pivoted floor springs are more effective where constant use is expected e.g. shops and office reception areas, etc.

Sliding doors

These are mainly used either to economise on space where it is not possible to swing a door, or for large openings which would be difficult to close off with swinging doors.

Folding doors

These are a combination of swinging and sliding doors. They can be used as either movable internal partitions to divide up large rooms, or alternatively as doors for large warehouses and showroom entrances.

Revolving doors

These may be used as main entrances to hotels, offices and shops, etc. They allow a fairly constant flow of traffic while reducing draughts and heat loss, although they are often used adjacent to swinging doors in order to accommodate extra traffic and goods etc. A specialist firm normally carries out the design, construction and installation of these doors.

Door hingeing and hanging

Door hanging is carried out after the plastering is complete, but normally before skirting and architraves are fixed. Speed and confidence in door hanging can be achieved by following the procedures, see Figure 6.69.

1. Measure height and width of door opening.

2. Refer to door schedule(s) and select the correct door.

3. Locate and mark the top and hanging side of the opening and door. For flush and fire-check doors these should have been marked by the manufacturer. When the hanging side is not shown on the drawing, the door should open into the room to provide maximum privacy, but not onto a light switch.

4. Cut off the horns (protective extensions on the top and bottom of each stile) on panelled doors. Flush doors will probably have protective pieces of timber or plastic on each corner. These need to be prised off.

5. Where plain door linings are used, tack temporary stops either side at mid height to stop the door falling through the opening. These should be set back from the edge of the lining by the door thickness.

6. 'Shoot' (plane to fit) in the hanging stile of door to fit the hanging side of the opening. This should be planed at a slight undercut angle to prevent binding.

7. Shoot the door to width. Allow a 2-mm joint all around between door and frame or lining. Many woodworkers use a two pence coin to check. The closing side will also require planing to a slight angle to allow it to close. This is termed a leading edge.

Measure opening

Refer to schedules

Select door

Light switch

Door opens into room

Hanging side marked

Top

Fix temporary stop to plain linings

Locate top and hanging side. Cut off horns or remove corner protection

Door held in wedged block or vee of stool when planing

Top — Shoot hanging style to fit

Top — Shoot to width

Top — Shoot top to fit

Top — Shoot or saw bottom, allow 6 mm

Door — Use a 2p coin to check for 2 mm joint

Leading edge required

Scribe bottom of door with pencil to fit threshold or carpet strip allow 2 mm

Top — Wedge in opening, mark hing positions

Mark and recess hinges Screw one leaf to door

Wedge door in place, secure with one screw per hinge

Use smoothing plane to remove arrises, adjust door fit if required

Fit remaining screws

▲ **Figure 6.69** Door hanging procedure

8. Shoot the top of the door to fit the head of the opening. Saw or shoot the bottom of the door to give a 6-mm gap at floor level or to fit the threshold.

9. External doors may require rebating along the bottom edge, to fit over the water bar.

10. Mark out and cut in the hinges (see later hinge recessing procedure). Screw one leaf of each hinge to the door.

11. Offer up the door to the opening and screw the other leaf of each hinge to the frame.

12. Adjust fit as required. Remove all arrises (sharp edges) to soften the corners and provide a better surface for the subsequent paint finish. If the closing edge rubs the frame, the hinges may be proud and require the recesses to be cut deeper. If the recesses are too deep, the door will not close fully and tend to spring open, which is known as 'hinge bound'. In this case a thin cardboard strip can be placed in the recess to pack out the hinge.

13. Fit and fix the lock or latch (see later lock fitting procedure). This correctly positions the closing edge of the door flush with the face of the doorframe or lining.

14. Remove temporary stops from plain linings and replace with planted stops pinned in position. Fix the head first, then the sides (Figure 6.70). Rectangular section stops can be simply butted; moulded sections will require mitring. Allow a gap of up to 2 mm between the face of door and the edge of the stop. Otherwise it will bind after painting.

15. Finally fit any other ironmongery, such as bolts, letter plates and handles, etc.

It is usual practice to only fit and not to fix the other ironmongery, such as handles and bolts, etc., at this stage. They should be fixed later during the finishing stages, after all painting works are completed and the building is more secure.

Butt joint rectangular stops

Fix stop to head first

Pin stops and punch heads below surface

Up to 2 mm gap between door and stop to allow for painting

Mitre moulded stops

◀ **Figure 6.70** Fixing door stops to plain linings

Weatherboard

Inward opening external doors, in exposed positions, may require a weatherboard (Figure 6.71). These are screwed to the bottom face of the door to throw rainwater clear of the water bar. The ends may be kept clear of the frame or let into it.

▲ **Figure 6.71** Weatherboard for inward-opening external doors

Hinge recessing

The leaves of a hinge can be recessed into the door and frame equally, termed half-and-half. See Figure 6.72. Alternatively the hinge can be offset, with the front edge of both leaves being recessed into the door leaving a clean unbroken joint line. This 'all-in the door' method is more popular for cabinet work than door hanging. The actual method used is dependent on personal preference as each is equally as good; check with the specification, foreman or customer.

▶**Figure 6.72** Hinge recessing and knuckle positioning

The position of the knuckle in relation to the face of the door is optional, but is sometimes dependent on the hinge. Full-size room doors are mainly hung with the full knuckle protruding beyond the door face, to give increased clearance.

Some people like to see the leaf recessed up to the centre of the knuckle for a neater appearance. However this is mainly used for cabinet doors. Once again check the specification and personal preferences.

Hinge recessing procedure

Follow Figure 6.73:

1. Decide on the hinge position. This will depend on the size and type of door being hung. (See 'Hinge size and positioning' for further details.)

2. Mark the hinge position on both the frame and the door or component to be hung.

3. Position the hinge on the edge of the door and mark the top and bottom of the leaf with a sharp pencil.

4. Repeat the process to mark the leaf position on the frame.

5. Set a marking gauge to the width of the hinge leaf and score a line on the edge of the door and frame, between the two pencil lines.

6. On frames and linings with a stuck rebated stop a combination square can be set to mark the leaf width with a pencil.

7. Reset the marking gauge to the thickness of the hinge leaf. Gauge the door face and the frame edge at each hinge position.

8. Use a chisel held vertically and mallet to chop the ends of each hinge housing position to the recess.

9. Use the chisel bevel side down and a mallet to feather each hinge housing to depth.

10. With the chisel bevel side up pare the feathered housing to the gauged depth. Use the chisel at a slight angle so that the bottom of each recess is slightly undercut towards the back.

11. Screw the hinges to the door first. Then offer up the door and screw to the frame.

▲ **Figure 6.73** Hinge recessing

When using brass screws and hinges, it is good practice to screw the hinges on initially with a matching set of steel screws. The brass screws being softer are easily damaged and may even snap off when being screwed in. The steel screws pre-cut a thread into the pilot holes for the brass screws to follow easily without risk of damage. Candle wax or petroleum jelly may be applied to lubricate the screw thread before insertion. If slot-head screws are being used, these should be lined up vertically. This gives an enhanced appearance and prevents the build up of paint or polish in the slots.

Door ironmongery

Door ironmongery is also termed **door furniture** and includes hinges, locks, latches, bolts, other security devices, handles and letter or postal plates. The actual items of ironmongery used will be determined by the method of operation, location, construction and desired performance requirements.

Door handing

The **hand** of a door must be established in order to select the correct items of ironmongery. Some locks and latches have reversible bolts, enabling either hand to be adapted to suit the situation. Traditionally, this has always been done by viewing the door from the hinge knuckle side; if the knuckles are on the left the door is left-handed, whereas if the knuckles are on the right, the door is right-handed. Doors may also be defined as either clockwise or anti-clockwise closing when viewed from the knuckle side. When ordering ironmongery simply stating left-hand or right-hand, clockwise or anti-clockwise can be confusing, as there may be variations between manufacturers and suppliers. The standard way now to identify handing is to use the following coding (see Figure 6.74):

- 5.0 for clockwise closing doors and indicating ironmongery fixed to the opening face (knuckle side).

- 5.1 for clockwise closing doors and indicating ironmongery fixed to the closing face (non-knuckle side).

- 6.0 for anti-clockwise closing doors and indicating ironmongery fixed to the opening face.

- 6.1 for anti-clockwise closing doors and indicating ironmongery fixed to the closing face.

▲ **Figure 6.74** Method for stating door handing

Hinge materials and types

Hinges are used to provide the opening and closing action required for most swinging doors. They are available in a variety of materials.

- *Pressed mild steel* hinges are the least expensive and are commonly used for internal painted doors. They are also available in different finishes such as electro-brass, bright zinc or chrome plated.

- *Solid brass and stainless steel* hinges are used for hardwood clear finished doors, both internally and externally. Steel hinges should not be used on hardwood or external doors, because of rusting and subsequent timber staining.

- *Cast iron* hinges are more suited to painted heavyweight doors. However, take care when fitting as they are brittle and easily broken if hit with a hammer.

- *Plastic and certain brass* hinges and other low melting point materials are not suitable for use on fire doors.

Always consult manufacturers' details regarding the suitability of hinges for a particular end use. Figure 6.75 illustrates a range of hinges, which are readily available in a variety of materials.

- *Butt hinge* is a general-purpose hinge suitable for most applications. It consists of two leaves, joined by a pin that passes through the knuckle formed on the edge of both leaves. As a general rule the leaf with the greatest number of knuckles is fixed to the doorframe.

- *Flush hinges* can be used for the same range of purposes as a butt hinge on both cabinets and full size room doors. They are only really suitable for lightweight doors, but they do have the advantage of easy fitting, as they do not require 'sinking in'.

- *Loose pin butt hinges* enable easy door removal by knocking out the pins. For security reasons they should not be used for outward opening external doors.

- *Lift-off butt hinges* also enable easy door removal, the door being lifted off when in the open position. These are available as right or left-handed pairs. Each pair consists of one long pin hinge, which is fitted as the lower hinge. The upper hinge has a slightly shorter pin, to aid repositioning the door.

- *Washered butt hinges* are used for heavier doors to reduce knuckle wear and prevent squeaking.

- *Parliament hinges* have wide leaves to extend knuckles and enable doors to fold back against the wall thus clearing deep architraves, etc.

- *Rising butts* are designed to lift the door as it opens to clear obstructions such as mats and rugs. They also give a door some degree of self-closing action. In order to prevent the top edge of the door fouling the frame as it opens and closes, the top edge must be eased. The hand of the door must be stated when ordering this item.

- *Soss hinges* are available in a range of sizes which depend on the weight of the door. They are fitted by drilling and recessing into the door and frame. Their main feature is that they are invisible when the door is closed.

▲ **Figure 6.75** Range of hinges for doors

- *Strap hinges* are surface fixed, being screwed or bolted directly to the door and frame or post.
- *Tee hinges* are used for ledged and braced doors and gates. Decorative versions are also available.
- *Hook and band hinges* are suitable for more heavy-duty use.
- *Double and single action spring hinges* are designed to make a door self-closing. Their large knuckles contain helical springs, which can be adjusted to give the required closing action by using a tommy bar and moving the small pin at the top of the knuckles into a different hole. Where double-action hinges are used, they should not be cut into the doorframe but screwed to a planted section the same width as the door and fixed to the frame.
- *Hawgood hinges* are another type of spring hinge and are suitable for industrial and heavy-duty double-swing doors. The spring housed in the cylinder is mortised and recessed into the doorframe, while the moving shoe fits around both sides of the door. A twin-spring Hawgood hinge is available for use as the top hinge of very heavyweight doors.

Hinge size and positioning

Lightweight internal doors are normally hung on one pair of 75-mm hinges; glazed, half-hour fire resistant and other heavy doors need one pair of 100-mm hinges. All external doors and one-hour fire resistant doors need three ($1\frac{1}{2}$ pairs 100-mm hinges. The standard hinge positions for doors are illustrated in Figure 6.76. For flush doors these are 150 mm down from the top, 225 mm up from the bottom and the third hinge where required, positioned centrally to prevent warping, or towards the top for maximum weight capacity. On panelled and glazed doors the hinges are often fixed in line with the rails to produce a more balanced look. However, slight adjustment may be required to avoid the end grain of wedges.

Pivoted floor springs

These provide a superior method of swinging and controlling both single- and double-action doors, although their use is restricted mainly to shops and offices etc. because of their expense.

▲ **Figure 6.76** Hinge positioning

Figure 6.77 illustrates single and double floor springs. Both consist of:

- A spring contained within a metal box, which is bedded into the floor screed.
- A shoe, which is fixed to the bottom of the door and locates over the pivot spindle.
- A two-part top pivot, of which one part is fixed to the head of the frame and the other to the door.

single action

double action

▲ **Figure 6.77** Pivoted floor springs

It is essential for the smooth operation of the door that the loose box and floor spring is set into the floor at the correct position to the frame (see Figure 6.78). This is determined by the finished thickness of the door plus an allowance between the door in its opening position and frame of between 3 and 6 mm.

The closing action of a door fitted with a floor spring is illustrated in Figure 6.79. A stand-open position, normally at 90°, can be included to hold the door in the open position, although the stand-open device is not suitable for fire doors.

The delayed closing action over the first few degrees can delay the closing of the door for up to one minute before it reverts to its normal

▲ **Figure 6.78** Floor spring loose box positioning

▲ **Figure 6.79** Action of door with floor spring

closing speed and is therefore suitable where goods and trolleys, etc., have to pass through. Turning the closing adjustment screw in the floor spring will vary the door's closing action.

The single-action doors may be removed from the frame by unscrewing the head of the frame plate, tilting the door slightly outwards to clear the frame and lifting it off the bottom pivot spindle.

Double-action doors are removed by first retracting the top pivot pin. This is achieved by turning the adjustment screw in the head of the frame plate. The door can then be tilted slightly outwards and lifted off the bottom pivot.

Brush or rubber seals may be required around the edges of double-action doors to prevent the passage of smoke and draughts. These may also include intumescent strips for fire doors (see Figure 6.80).

▲ **Figure 6.80** Double action door seals

Door closers

These are mainly used in offices, shops and industrial premises to give these normally larger and heavier doors a controlled self-closing action, and also to enable the door to be held in the open position when required (Figure 6.81). Turning the adjustment screw, which is normally under the main cover, will alter the speed of closing. Certain types of closer are fitted with a temperature-sensitive device that automatically closes the door from the held-open position in the event of a fire.

Other door closers are available with the following actions:

- Delayed closing action as with floor springs.
- Snap action over the last few degrees of travel to ensure that the door closes fully into the latch.
- Back check action this controls the last few degrees of opening travel, thus preventing the door from slamming into a return wall adjacent to the opening.

▲ **Figure 6.81** Overhead door closure

193

▲ **Figure 6.82** Concealed door closure

A **concealed spring door closer** is illustrated in Figure 6.82. The cylinder which contains the spring is mortised into the edge of the door and the plate recessed flush. The anchor plate is recessed in and screwed to the frame. Adjustment to the closing action is achieved by inserting the metal plate near the cylinder plate to hold the chain, unscrewing the anchor plate and turning it either clockwise to speed the action or anticlockwise to slow the closing action.

The **hydraulic door check and holder** shown in Figure 6.83 can be used in conjunction with door closers or spring hinges to control the last few degrees of closing travel. This prevents slamming and holds the door firmly closed against strong winds.

In order to ensure that self-closing double doors with rebated stiles close in the correct sequence, a **door selector** must be fitted (see Figure 6.84).

▲ **Figure 6.84** Double-door selector

▲ **Figure 6.83** Hydraulic door check and holder

Locks, latches and other door furniture

Always consult manufacturers' details regarding the suitability of locks, latches and other door furniture for a particular end use. Figure 6.85 illustrates a range of commonly used items, which are readily available in a variety of materials and finishes.

See Table 6.1 for typical materials used to manufacture ironmongery and their common abbreviations.

Locks and latches

■ *Cylinder rim night latches.* Mainly used for entrance doors to domestic property, but as they are only a latch, provide little security on their own. When fitted, the door can be opened from the outside with the use of a key and from the inside by turning the handle. Some types have a double locking facility that improves their security. Double locking types are essential when used for glazed

Cylinder rim latch

Mortise deadlock

Mortise latch

Digital keypad

Knob

Barrel bolt

Tower bolt

Neck bolt

Mortise lock/latches

Horizontal

Vertical

Flush bolt

Mortise rack bolt

Euro pattern

Cylinder lock

Rebated mortise lock/latch

Hinge bolt

Rim deadlock

Rim lock and latch

Mortise latch

Knobset

Keyhole escutcheon

Knob furniture

Euro lever combined or with separate escutcheon

Inside turn button

Bathroom privacy lever combined or with separate estutcheon

Outside screw slot

Monkey tail bolt

Bow 'D' handle bolt

Lever furniture

Security chain

Thumb latch

Panic bolts

Alarm

Push bar

Push latch

Door holder

Padlock, hasp and staple

▲ **Figure 6.85** Locks, latches and other door furniture

Table 6.1 Ironmongery finishes

Abbreviation	Material finish
Al	Aluminium
An Al	Anodised aluminium
B	Brass
BJ	Black japanned
BMA	Bronze metal antique
CP	Chrome plated
EB	Electroplated brass
NP	Nickel plated
PA	Polished aluminium
PB	Polished brass
PC	Powder coated
PNP	Polished nickel plated
SAA	Satin anodised aluminium
SB	Satin brass
SC	Satin chrome
SS	Stainless steel
ZP	Zinc plated

doors, as the inside handle will not turn even if an intruder breaks a pane to reach inside for it.

■ *Mortise deadlocks.* Provide a straightforward key-operated locking action and is often used to provide additional security on entrance doors where cylinder rim latches are fitted. They are also used on doors where simple security is required, e.g. storerooms. The more levers a lock has, the better it is. A 5-lever lock is more difficult to pick than a 3-lever one. For public buildings, shops and offices, a *master keyed system* is available. The 'grand master key' will operate all locks in a system and sub-masters will operate only a specific range of locks in a system. 'Servant' keys will only operate individual locks.

■ *Mortise latches* are used mainly for internal doors that do not require locking. The latch, which holds the door in the closed position, can be operated from either side of the door by turning the handle.

■ *Mortise lock/latches* are available in two main types. The horizontal one is little used nowadays because of its length, which means that it can only be fitted to doors with substantial stiles. The vertical type is more modern and can be fitted to most types of doors. It is often known as a narrow-stile lock/latch. Both types can be used for a wide range of general-purpose doors in various locations. They are, in essence, a combination of the mortise deadlock and the mortise latch. A further variation is the Euro pattern mortise lock/latch, which uses a cylinder lock to operate the dead bolt. Also, bathroom privacy locks are available which use a turn button on the inside of the door.

■ *Rebated mortise lock/latches* should be used when fixing a lock/latch in double doors that have rebated stiles. The front end of this lock is cranked to fit the rebate on the stiles. A conversion kit is also available for use with a standard mortise lock/latch.

■ *Rim deadlocks* are surface fixed locks, traditionally used in the place of a mortise deadlock, now rarely used.

■ *Rim lock/latches* are surface fixed versions of the mortise lock/latch. Mainly used today for garden gates and sheds. Knob furniture is used to operate the latch.

Lock accessories

■ *Knobsets.* These consist of a small mortise latch and a pair of knob handles that can be locked with a key so that it can be used as a lock/latch in most situations both internally and externally. Knobsets can also be obtained without the lock in the knob for use as a latch only.

■ *Knob furniture* is for use with rim and horizontal mortise lock/latches. It should not be used with vertical mortise lock/latches, as the knob is very close to the edge of the door, hand injuries may be caused when closing the door (from the closing face). Versions are available with a digital key pad *entry system* for controlled access.

■ *Keyhole escutcheon plates* are used to provide a neat finish to the keyhole of both deadlocks and horizontal mortise lock/latches. A bathroom privacy version has a screw slot on the outside for use in an emergency.

- *Lever furniture* is available in a wide range of patterns, for use with the mortise latches and mortise lock/latches. A bathroom privacy version has a screw slot on the outside for use in an emergency

- *Thumb latches* are also known as a Norfolk or Suffolk latches. They are used to latch doors and gates in the closed position and may be installed either side of the door.

- *Padlock, hasp and staple.* Used on external doors such as sheds and gates to provide security. The hasp is fixed to the door and the staple to the frame. Larger versions are fixed by bolting through the door rather than screwing, to provide greater protection.

Bolts and chains

- *Barrel and tower bolts* are used on external doors and gates to secure them from the inside. Two bolts are normally used, one at the top of the door and the other at the bottom. Heavier versions such as the monkey tail bolt and the D-handle bolt are used for garage and industrial doors.

- *Flush bolts* are flush fitting and therefore require recessing into the timber. They are used for better quality work on the inside or edge of external doors to provide additional security and also on double doors and French windows to bolt one door in the closed position. Two bolts are normally used, one at the top of the door and the other at the bottom.

- *Mortise rack bolts* are a fluted key operated dead bolt, which is mortised into the edge of a door at about 150 mm from the top and bottom.

- *Hinge bolts* help to prevent a door being forced off its hinges, particularly on outward opening external doors where the hinge knuckle pin is vulnerable.

- *Security chains* can be fixed on front entrance doors, with the slide to the door and the chain to the frame. When the chain is inserted into the slide, the door will only open a limited amount until the identity of the caller is checked.

- *Panic bolts and latches* are used on the inside of emergency exit doors to bolt them closed. Pressure applied to the push bar will disengage the bolts. They can be fitted with an audible alarm to deter unauthorised use.

- *Door holders* are foot-operated door stops to hold doors and gates in the open position. They are particularly useful for holding side-hung garage doors open while driving into or out of the garage.

Positioning locks, latches and door furniture

The position of door ironmongery or furniture will depend on the type of door construction, the specification and the door manufacturer's instructions. Figure 6.86 illustrates the recommended fixing height for various items.

- The standard position for mortise locks and latches is 990 mm from the bottom of the door to the centreline of the lever or knob furniture spindle. However, on a panelled door with a middle rail, locks/latches may be positioned centrally in the rail's width.

Mortise locks
and latches

Cylinder rim latch
and letter plate

Typical external door
ironmongery

Plates
and signs

▶**Figure 6.86** Fixing heights

- Cylinder rim latches are positioned in the door's stile between 1200 mm and 1500 mm from the bottom of the door and the centre-line of the cylinder.

- Before fitting any locks/latches the width of the door stile should be measured to ensure the lock/latch length is shorter than the stile's width, otherwise a narrow stile lock may be required.

- On external doors using both a cylinder rim latch and a mortise dead lock, the best positions for security is one third up from the bottom for the dead lock and one third down from the top for the cylinder rim latch.

- Security chains are best positioned near the centre of the door in height.

- Hinge bolts should be positioned just below the top hinge and just above the bottom hinge.

Miscellaneous door furniture

- *Letter plates* are normally positioned centrally in a door's width and between 760 mm and 1450 mm from the bottom of the door to the centre line of the plate. Again on panelled doors letter plates may be positioned centrally in a rail and sometimes even vertically in a stile.

- *Information signs* are normally positioned centrally in a door's width and 1500 mm from the bottom of the door to the centre of the sign.

- *Kicking plates* are fixed to the bottom face of the door for protection, keeping an even margin. *Push or finger plates* are positioned near the closing edge of a door at a height of 1200 mm to the centre of the plate. On panelled and glazed doors, they should be fixed centrally in the stile's width.

Before starting work, always read the job specification carefully as exact furniture positions may be stated. Also read both the door manufacturer's and the ironmongery manufacturer's instructions to ensure the intended position is suitable to receive the item, e.g., the positioning of the lock block on a flush door, and the correct fixing procedure.

Sliding door ironmongery

Various types of sliding gear are available for straight sliding, bypassing and around-the-corner arrangements. In general, top-sliding gear is used for lightweight doors and bottom-sliding gear is used for heavier doors.

Figure 6.87 shows typical details and sections of an internal sliding door which is top hung. The bottom of the door is controlled by a small

◀ **Figure 6.87** Internal sliding door

nylon guide that runs in a channel set in the bottom of the door. Sliding doors require special locks and latches with hook-shaped bolts, which prevent the door being slid open. The furniture is usually flush fitting so that the doors are able to slide without fouling.

Folding door ironmongery

Folding doors may be either end folding or centre folding. Top-hung or bottom-roller gear is available for both types.

End-folding units should consist of an even number of leaves, up to six, hinged on either jamb of the frame (Figure 6.88). For extra wide openings, additional 'floating' units of four or six leaves not hinged to either jamb may be added. An extra leaf can be added if a swinging access door is required. A leaf width of between 600 mm and 900 mm is recommended. These must have solid top and bottom rails so that a firm fixing for the edge-fitting hangers and guides can be achieved. Figure 6.88 shows the arrangement and sections of an external end-folding top-hung unit.

Centre-folding units are of a similar construction to edge-folding units except that there will always be a half-leaf hinged to the frame to permit the centre folding arrangement. From $1\frac{1}{2}$ to $7\frac{1}{2}$ leaves may be

▲ **Figure 6.88** End-folding doors

hinged together on either jamb of the frame. An extra leaf can be added if a swinging access door is required. Figure 6.89 shows details and sections of an internal centre-folding bottom-running partition.

Installation of mortise dead lock, latch or lock/latch

See Figure 6.90. Always consult manufacturers' instructions.

1. Wedge the door in the open position. Use the lock as a guide to mark the mortise position on the door edge at the pre-marked height.

2. Set a marking gauge to half the thickness of the door. Score a centre line down the mortise position to mark for drilling.

3. Select a drill bit the same diameter as the thickness of the lock body. An oversize bit will weaken the door; use an undersize one and additional paring will be required. A hand brace and bit or power drill and spade bit may be used. Mark the required drilling depth on the bit using a piece of masking tape. Drilling too deep will again weaken the door as well as the risk in panel and glazed doors of the drill breaking out right through the stile.

4. Drill out the mortise working from the top, each hole slightly overlapping the one before.

5. Use a chisel to pare away waste between holes to form a neat rectangular mortise.

6. Slide the lock into the mortise and mark around the faceplate.

7. Remove the lock and use a marking gauge to score deeper lines along the grain, as this helps to prevent the fine edge breaking out or splitting when chiselling out the housing.

8. Use a chisel to form the housing for the faceplate (let in).

9. The surface can first be feathered (as when recessing hinges) and finally cleaned.

10. Hold the lock against the face of the door, with the faceplate flush with the door edge and lined up with the faceplate housing. Mark the centre positions of the spindle and keyhole as required with a bradawl.

11. Use a 16-mm drill bit to drill the spindle hole, working from both sides to avoid breakout, or clamp a waste piece to the back of the door.

12. Use a 10-mm drill bit to drill the keyhole, again working from both sides. Cut the key slot with a padsaw and clean out with a chisel to form a key guide Alternatively use a 6-mm drill bit to drill out a second hole below the 10-mm hole and chisel out waste to form a key guide Never drill a larger hole for the key as it will not give a guide when inserting the key, making it a hit-and-miss affair.

13. Insert the lock, check the spindle and keyholes, and align from both sides. Secure the faceplate with screws. Re-check that the key works.

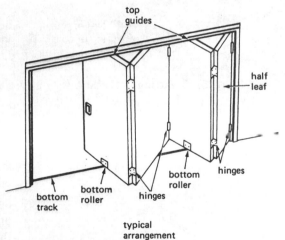

▲ **Figure 6.89** Centre-folding doors

Figure 6.90 Fitting procedure for mortise locks and latches

Wedge door open

Mark position on door edge

Gauge centre line on door edge

Drill out to width and depth of lock

Pare sides to form mortise

Mark lock face plate

Let-in face plate

Mark spindle and keyhole centres as required

Drill holes, cut keyhole to guide key

Drill hole and clean out guide

Drill two holes and clean out between

Never one large hole

Mark bolt position on frame/lining

Mark position of striking plate

Let-in striking plate, cut mortise for bolts

▲ **Figure 6.90** Fitting procedure for mortise locks and latches

14. With the dead bolt out, close the door against the frame and mark the bolt and latch positions on the edge of the frame. Square these positions across the face of the frame.

15. Set the adjustable square from the face of the door to the front edge of the latch or dead bolt. Use it to mark the position on the face of the frame.

16. Hold the striking plate over the latch or dead bolt position and mark around the striking plate. Gauge vertical lines to prevent breakout when chiselling.

17. Chisel out to let in the striking plate. Again you may find it easier to feather first before finally cleaning out. The extended lead-in or lip for the latch may require a slightly deeper bevelled housing or recess.

18. Check for fit and screw the striking plate in place. Select a chisel slightly smaller than the striking plate bolt-holes.

19. Chop mortises to accommodate both the latch and dead bolt. Some striking plates have boxed bolt-holes. These must be cut beforehand.

20. Finally fit the lever furniture, knob furniture or keyhole escutcheon plates as appropriate and check for smooth operation. When fixing keyhole escutcheon plates, the key should be passed through the plate and into the lock and centralised on the key shaft before screwing.

21. Finally fit the lever furniture, knob furniture or keyhole escutcheon plates as appropriate and check for smooth operation.

Installation of a cylinder rim lock

See Figure 6.91. Always consult manufacturers' instructions.

1. Wedge the door in an open position. Use the template supplied with the lock at the pre-marked height to mark the centre of the cylinder hole.

2. Use a 32-mm auger bit in a brace or a spade bit in a power drill to drill the cylinder hole. Drill from one side until the point just protrudes. Complete the hole from the other side to make a neat hole, avoiding breakout, or alternatively by cramping a block to the door.

3. Pass the cylinder through the hole from the outside face and secure it to the mounting plate on the inside with the connecting machine screws.

4. Ensure that the cylinder key slot is vertical before fully tightening the screws. For some thinner doors these machine screws may require shortening before use with a hacksaw. If required, secure the mounting plate to the door with woodscrews.

5. Check the projection of the flat connection strip. This operates the latch and is designed to be cut to suit the door thickness. If necessary use a hacksaw to trim the strip so that it projects about 15 mm past the mounting plate.

6. Align the arrows on the backplate of the rim lock and the turnable thimble.

7. Place the rim lock case over the mounting plate ensuring that the connection strip enters the thimble.

8. Mark out and let in the rim lock lip in the edge of the door if required.

9. Secure the rim lock case to the door or mounting plate with wood or machine screws as required. Check both the key and inside handle for smooth operation.

10. Close the door and use the rim lock case to mark the position of the keep (striking plate) on the edge of the doorframe.

11. Open the door and use the keep to mark the lip recess on the face of the frame. Chisel out a recess to accommodate the keep's lip. Secure the keep to the frame using woodscrews.

12. Finally check from both sides to ensure a smooth operation.

Use template supplied to mark centre for hole

Drill hole, insert cylinder, let-in rimlock if required

Cut connection strip to length if required

With rimlock screwed in place mark striking plate position

Let-in striking plate, screw in place

▲ **Figure 6.91** Fixing a cylinder rim night latch

Installation of a letter plate

See Figure 6.92. Always consult manufacturers' instructions.

1. Wedge the door in the open position. Mark the centreline of the plate on the face of the door.
2. Position the plate over the centreline and mark around it.
3. Measure the size of the opening flap and mark the cut-out on the door. Allow about 2 mm larger than the flap, to ensure ease of operation.
4. Mark the position of the holes for the securing bolts.
5. If the door is easily removed, cramp an off-cut of timber to the back of the door to prevent damage from drill breakout. Alternatively the holes can be drilled from both sides.
6. Drill holes for the fixing bolts and at each corner of the flap cut-out.
7. Use a jigsaw or padsaw to saw from hole to hole.
8. Neaten up the cut-out if required using glass paper. Remove the arris from the inside edges.
9. Position the letter plate and secure, using the fixing bolts.
10. Check the flap for ease of operation and adjust if required.

▶**Figure 6.92** Installing a letter plate

Installation of barrel bolts

See Figure 6.93. Always consult manufacturers' instructions.

1. Place the bolt in the required position. Mark one of the screw holes through the backplate with a bradawl or pilot drill.
2. Insert a screw, ensuring the bolt is parallel or square to the edge of the door, and insert a screw at the other end of the bolt.
3. Move the bolt to the locked position and slide the keep over the bolt.
4. Mark the screw holes in the keep and screw in place.
5. Check the bolt works smoothly before inserting the remaining screws.

Cranked or swan necked bolts will require a hole drilling in the head of the doorframe to receive the bolt:

1. Position and secure the bolt as before.
2. Slide the bolt to the locked position and mark around the bolt.

Square up, insert second screw

Insert first screw

Position and screw on keep

Insert remaining screws

Keep plate may be let in or surface fixed

Swan neck bolts will require a hole in the head of the frame to receive the bolt

▲ **Figure 6.93** Installing barrel bolts

3. Use an auger bit slightly larger than the width of the bolt to drill a hole in the marked position. Ensure the bolt can slide to its full length.

4. Check for smooth operation.

Installation of a mortise rack bolt

See Figure 6.94. Always consult manufacturers' instructions.

1. Use a marking gauge to score a centreline on the edge of the door.

2. Mark the centre of the bolt-hole using a try square and pencil. Transfer the line onto the face of the door that the bolt will operate from.

3. Select an auger or spade bit slightly larger than the bolt barrel. Drill a hole in the edge of the door. Use a piece of tape wrapped around the bit as a depth guide.

4. Insert the bolt, and then turn the faceplate so that it is parallel to the door edge. Mark around the faceplate with a pencil.

5. Remove the bolt and gauge the parallel edges and using a sharp chisel 'let in' the faceplate.

6. Place the bolt on the face of the door. Use the bradawl to mark the key position.

7. Use a 10-mm bit to drill the keyhole through the face of the door, into the bolt-hole.

8. Insert the bolt into the hole then check that the faceplate finishes flush and that the keyhole lines up. Insert the fixing screws.

9. Insert the key through the keyhole plate into the bolt. Position the keyhole plate centrally over the key and insert the fixing screws.

10. Close the door and turn the key to rack the bolt to its locked position. The point on the bolt will mark the clearance hole centre position on the frame.

Gauge centre line on door edge

Mark centre line on edge and face of door

Tape

Select auger or spade bit to match bolt. Use tape to mark required depth

Drill hole in door edge

Insert bolt and mark around face plate

Let in face plate

Place bolt on face of door and mark key position

Bolt point will mark centre on frame

Turn key to rack bolt

▲ **Figure 6.94** Installing a mortise rack bolt

11. Drill the clearance hole for the bolt in the frame on the marked centre.

12. Close the door. Rack the bolt and check for the correct alignment of the bolt clearance hole.

13. 'Let in' the keep centrally over the clearance hole and insert the fixing screws.

14. Check for smooth operation.

Doorframes and linings

Doors can be hung on frames or linings, which have been built-in, or fixed into an opening in a wall or partition.

Standard doorframes

Doorframes consist of a head, two jambs and, where required for external use, a sill or threshold (Figure 6.95). These component parts are jointed together with draw-pinned mortise and tenon joints. The threshold should have a galvanised metal or plastic water bar fitted to prevent the passage of driving rain under the door. The door will have to be rebated along its bottom edge to fit over the bar. As an additional protection against rainwater running down the face of the door and underneath, a weatherboard can be fitted to throw it clear.

▲ **Figure 6.95** Door frame details

Hole in tenon offset towards the shoulder

Head

Jamb

Hammer in pointed dowel to pull up shoulder

Head

Jamb

Saw a 'Vee' in end of dowel, drive-in wedge

Trim-off wedged dowel flush with surface

▲ **Figure 6.96** Draw pinning (joint cut away to show detail)

Draw pins are used in preference to wedges, as they will hold the joint even if the horn is cut off. In addition the use of draw pins has the advantage of not requiring cramps to pull up the joint during assembly. By off-setting the hole in the tenon slightly towards the shoulder, the joint will be drawn up tight as the pin is driven in (see Figure 6.96). The draw pin can simply be cut off flush with the frame after driving in, or have a wedge added.

Storey height doorframes

These are used for door openings in thin non-load-bearing blockwork partitions (Figure 6.97). The jambs and head, which make up the frame, are grooved out on their back face to receive the building blocks. A mortise and tenon joint is used between the head and the jambs. The jambs above the head are cut back to finish flush with the blockwork. These should be covered with expanded metal before plastering.

Storey height doorframes with fanlight

Figure 6.98 shows a frame for internal use, and Figure 6.99 a frame for external use. Both types consist of two jambs, a head and a transom. The external frame also has a hardwood sill with a galvanised water bar, which has been bedded into a plough groove with mastic. The joints used for both frames are mortise and tenons.

head

jamb

pinned door stop

transom

head

mortise and tenon

▲ **Figure 6.97** Internal storey height frame and joint detail

head

glazing beads

transom

jamb

pinned door stop

sill

◀ **Figure 6.98** Storey height frame with fanlight for internal use

◀ **Figure 6.99** Storey height frame with fanlight for external use

▲ **Figure 6.100** Internal storey height transom joints

▲ **Figure 6.101** Transom joint for external storey height frame with fanlight

For a plain internal frame the joint between the transom and jamb can be a single mortise and tenon but where a rebated frame is used, the joint should be a double mortise and tenon as shown in Figure 6.100.

On external storey height frames where the transom extends beyond the face, the joint will also be a double mortise and tenon and the overhanging edge of the transom should be housed across the face of the jambs for improved durability (see Figure 6.101).

Other doorframes, linings, screens and lobbies

Curved head frames

It is normal to form the head of curved doorframes from solid sections, which are jointed together to produce the required shape. Alternatively, the curved member may be built up using a number of layers with end joints staggered and glued together (see Figure 6.102).

▲ **Figure 6.102** Forming curved members

▲ **Figure 6.103** Curved head door frames

A third method of forming the curved head is by glulam construction. However, this is not often used as it is the most expensive, requires a cramping jig and the grain pattern resulting from the thin laminations may not be suitable for high quality work.

The shape of the head is used to classify curved head doorframes. Four typical shapes, along with suitable jointing methods are illustrated in Figure 6.103.

The main methods used to join the members are either bridle joints and draw pins, or handrail bolts and dowels (see Figure 6.104).
Where a transom is required near the springing point, it should be jointed to the jamb using two tenons in order to minimise short graining problems. Hammer-headed tenons and keys could also be used to join the curved members together, although they are less popular, as they are not suitable for machine production methods.

Linings

Plain linings consist of two plain jambs and a plain head joined together using a housing or a bare-faced tongue and housing for better quality work. The planted stop is fixed around the lining after the door has been hung. (See Figure 6.105.)

Rebated linings or door casings are used for better quality work. They consist of two rebated jambs and a rebated head. The rebate must be slightly wider than the width of the door so that when the door is hung it will finish flush with the edges of the lining, but still have a working clearance of 1.5 mm between the door and the stop in order to allow for finishing and prevent the door from binding or rubbing. (See Figure 6.106.)

head

draw
pin

mitred bridle

jamb

bridle

head

draw
pins

transom

jamb

two tenons

head

handrail bolt

dowel

jamb

▲ **Figure 6.104** Curved head frame joints

Head

Jamb

Pinned door stop

▲ **Figure 6.105** Plain lining

Head

Jamb

▶**Figure 6.106** Rebated lining

Skeleton linings are used for deeper reveals, where the brickwork is too thick for a normal lining to be used. They consist of a basic framework which is stub tenoned together. A stopped bare-faced tongue and housing is used between the head and the jambs. A ply or solid timber lining is used to cover the framework and form the rebate to receive the door. (See Figure 6.107.)

▶**Figure 6.107** Skeleton lining

Panelled linings (Figure 6.108) are normally used in conjunction with wall panelling. Their construction is similar to skeleton linings except that the framed jambs and head are grooved out to receive the panels and the stop is normally stuck-moulded on the solid as in Figure 6.95.

Splayed linings (Figure 6.109) are mainly used in conjunction with a door frame to finish deep reveals. They give a lead into the opening and create a distinctive feature. The linings, which may be plain, rebated, skeleton or panelled, are housed or mitred together at their head and tongued into the frame. In order to construct the lining, the true shape and bevels will have to be geometrically developed (see Book 1).

Door sets

Door sets consist of a pre-hung door and frame or lining, complete with all ironmongery (Figure 6.110). Loose pin or lift-off hinges are normally used for easy door removal. Architraves are normally fixed to one side of the set and loose pinned on the other. All work is undertaken in the joinery shop and supplied to site ready for fixing into its opening. The fixing of door sets should not be carried out until at least the second fixing stage. This is to avoid the possibility of any damage during the building process. The main advantages of using door-sets are the 'joiners shop quality' and the considerable saving in on-site time to install.

▲ **Figure 6.108** Panelled lining

▲ **Figure 6.109** Splayed lining

◀ **Figure 6.110** Door set

Glazed screens

The elevation of a glazed screen that incorporates a double-action door is illustrated in Figure 6.111. This type of frame is sometimes called a combination frame as it incorporates both door and glazing. Although the illustration shows a door, this could be omitted and the same detail used to construct a glazed partition.

The middle rail of the door has been continued across the side screen (also known as a side light). This continues the feature and also serves to break up the area of glass and provide protection at a vulnerable point.

The general arrangement of members is shown in the vertical and horizontal sections (see Figures 6.112 and 6.113). When the frame is fixed the mullion and plinth on the side screen must be secured with the aid of galvanised steel dowels grouted into the floor.

▲ **Figure 6.111** Glazed screen with door

horizontal section

▲ **Figure 6.112** Glazed screen horizontal section

Entrance lobbies

Traditionally it was fairly common for a lobby or vestibule to be built at the entrance to public buildings, banks and offices etc. Figure 6.114 shows a plan of such a lobby. It consists of a pair of single action entrance doors leading to a pair of double-action doors. The entrance doors are only intended to be closed when the building is locked up out of business hours, so they are often referred to as night doors. During the day they are kept open to give the impression of a panelled lobby. A pictorial sketch of this lobby with the entrance doors in the open position is shown in Figure 6.115. The actual walls of the lobby may be constructed in various ways, including brickwork, as shown in the plan, or timber framing.

Assembly of doorframes and linings

It is standard practice to assemble doorframes in the joiners shop, whereas linings are normally sent to site in flat-pack form, to be assembled by the carpenter.

Frames without a threshold are assembled with a temporary distance piece and braces to hold the jambs parallel and square. All frames are

head

glazing
bead

transom

top
rail

middle
rail

sill

bottom
rail

plinth

door

screen

steel dowel

vertical sections

◀ **Figure 6.113** Glazed screen
vertical section

normally assembled with temporary braces between the jamb and head
for squaring purposes. Both are illustrated in Figure 6.116.

Fixing of doorframes

Doorframes can be either 'built in' or 'fixed in':

- Built-in frames are fixed into a wall or other element by bedding in
  mortar and surrounded with the walling components.

▶**Figure 6.114** Plan
of entrance lobby

main area

swing of
doors

double hung
doors

position of night
door when
open

lobby

outside

entrance or
night doors

▶**Figure 6.115**
Entrance lobby sketch

Draw pinned joints no cramping required

Temporary braces pinned into rebate to hold frame square

Level bench bearers or stools

Temporary distance piece pinned into rebate to hold frames without threshold parallel

Apply waterproof glue to both exposed faces of tenon and shoulders

Head

Jamb

Bench bearer or stool

Tenon just inserted

▲ **Figure 6.116** Door frame assembly

■ Fixed-in frames are inserted into a ready-formed opening after the main building process.

Built-in frames

The bricklayer fixes the majority of frames by building in as the brickwork proceeds. Prior to this the frame has to be prepared, accurately positioned, plumbed, levelled and temporary strutted by the carpenter. Temporary struts are used to hold the frame upright. The feet of the doorframe jambs, in the absence of a threshold, are held in position by galvanised metal dowels, which are drilled into the end of the jambs and are grouted into the concrete. This is shown in Figure 6.117. Temporary braces and distance pieces are fixed to the frame in order to keep it square and the jambs parallel during the building-in process.

Temporary braces

Head   Horn

Jambs

Temporary strut

Bricks

Temporary distance piece   Galvanized metal dowels

▲ **Figure 6.117** Building in a frame

Level head

Temporary strut

Plumb and strut one jamb

Bed threshold level with mortar

Sight jambs for parallel

▶**Figure 6.118** Building in a frame with threshold

Galvanised frame cramp

▲ **Figure 6.119** Attaching frame cramps

Head

▲ **Figure 6.120** Horn cut back ready for building in

The frame's head should be checked for level and packed up as required. Frames with thresholds are normally bedded level using bricklayer's mortar (see Figure 6.118).

Jambs should be plumbed from the face. It is standard practice to plumb and strut the first using a spirit level. The other is then sighted parallel: stand to the side of the frame, close one eye, sight the edge of the plumbed jamb with the edge of the other and adjust if required until both jambs are parallel.

As the brickwork proceeds, galvanised metal frame cramps, as shown in Figure 6.119, should be screwed to the back of the jambs and built into the brickwork. Four or five cramps should be evenly spaced up each jamb.

The horns of the frame should be cut back as shown before building in, rather than cut off flush (see Figure 6.120). The horns will then help to fix the frame in position. After trimming the horns it is essential that the cut ends are treated with paint or preservative in order to reduce the possibility of timber rot.

Storey height frames may be used for internal door openings in thin blockwork partitions. The jambs and head that make up the frame are grooved out on their back face to receive the building blocks. The storey frame should be fixed in position – at the bottom to the wall plate and at the top of the joists – before the blocks are built up (see Figure 6.121). The jambs above the head are cut back to finish flush with the blockwork. As with other frames one jamb should be fixed plumb and the other sighted to it.

Fixed-in frames

These are installed into prepared openings in the wall. This applies mainly to expensive hardwood frames to protect them from possible damage or discoloration during the building process. In addition, frames that were not available during the building process or replacement frames will have to be fixed in.

Horns on fixed-in frames are not required as a fixing and should be sawn off flush with the back of the jambs. Remember to paint or preservative treat the cut ends.

Place the frame in the prepared opening, temporarily holding it in position with the aid of folding wedges (Figure 6.122). Check the head or sill for level and adjust the wedges as required.

▲ **Figure 6.121** Building in a storey height frame

▶**Figure 6.122** Positioning fixed-in frames

Plumb one jamb and 'sight in' the other; adjust the wedges if required.

Figure 6.123 shows how to fix the frame to the wall:

■ *Nailing* using cut nails into the blockwork or brickwork mortar joints or masonry nails into the actual brick.

■ *Screwing plastic plugs and screws through the jamb.* Screw heads in softwood frames may be countersunk below the surface and filled. Screw heads in hardwood frames should be concealed by counter boring and pelleting.

■ *Metal plates used as fixing lugs* screwed at intervals to the back of the jamb before the frame is put in the opening. The lugs are screwed and plugged to the brick or block reveals.

■ *Frame anchors* using proprietary fixing consisting of a metal or plastic sleeve and matching screw. The jamb and reveal are drilled out to suit the sleeve, which is inserted in position and screwed up tight.

▶**Figure 6.123** Methods of fixing frames

Fixing door linings

Door linings are normally fixed in place. The opening in the wall to receive the lining being formed while the wall is being built and the lining is fixed at a later stage. Fixings may be:

- *Nailing to twisted wooden plugs* (see 'Sequence of operations' below).

- *Nailing or screwing to timber pallets* that have been built into the brick joints by the bricklayer, or into the door stud of a stud partition. Folding wedges are used as packings down the sides of the jambs.

- *Nailing directly into the blockwork reveal or brickwork mortar joint.* Folding wedges will be required as packing.

- *Using plugs and screws* or other proprietary fixing.

Sequence of operations to fix a lining (using twisted timber plugs) as shown in Figure 6.124:

1. Assemble lining. This is normally done by skew nailing through the head into the jambs.

2. Fix a distance piece near the bottom of the jambs, and when required, diagonal brace or braces at the head.

3. Rake out brickwork joints and plug (see Figure 6.125). There should be at least four fixing points per jamb. Omit this stage if the bricklayer has built in wooden pallets or pads into the brickwork.

4. Offer lining into opening and mark where the plugs need to be trimmed. The plugs should project equally from both reveals.

5. Cut the plugs and check the distance with a width rod. The ends of the plugs should be in vertical alignment. Check with a straightedge and spirit level.

▲ **Figure 6.124** Fixing a lining

▲ **Figure 6.125** Plugs or pads used for fixing lining

6. Fix lining plumb and central in the opening by nailing or screwing through the jambs into the plugs. Before finally fixing check head for level, wedge off lintel, ensure the lining is out of wind: check by sighting through the jambs. When fixing to unplastered walls, check adjacent linings and wall surfaces are lined up.

7. Ensure lining jambs are packed up off a concrete sub-floor if required to suit the finished floor level (FFL).

7

Chapter Seven

Windows

This chapter covers the work of both the site carpenter and bench joiner. It is concerned with principles of constructing and installing windows. It includes the following:

- ❧ classification of windows
- ❧ the performance requirements of windows
- ❧ methods of window construction
- ❧ installation of windows
- ❧ types of window ironmongery
- ❧ single and double glazing
- ❧ fixing of windows and window boards
- ❧ window ventilation.

Windows are glazed areas in walls and roofs to allow natural lighting into buildings and to give the occupants an outside view. In addition they may also be used to provide rapid natural ventilation. They consist of glass panes held into a timber, metal or plastic frame, which in turn is fixed to the wall or roof structure.

The frame material is used as the first classification of windows e.g. timber or plastic windows, followed by whether or not they open. Windows that do not open are termed fixed or direct glazing; others are termed by their method of opening and whether or not they project from the main building line.

General classification of windows

The majority of windows come under one more of the following groups, which are illustrated in Figure 7.1.

- **Fixed glazing.** Glazing that is fixed directly into an opening or sub frame; also termed **direct glazing**.

- **Casements.** Normally opening windows that are either top or side, hung on hinges to a main frame, but may also be fixed.

- **Pivot hung windows.** A type of casement that is either horizontally or vertically hung and includes what is termed tilt-and-turn windows.

▲ **Figure 7.1** Classification of windows

- **Sliding sashes.** Opening windows which slide in a main frame either horizontally or vertically.

- **Bay windows.** These project beyond the main building line, often giving an increased floor area in the room. Bays which project from an upper storey only are known as **oriel windows**.

Design features

The design of windows requires careful consideration. In addition to the main functions, they have other performance requirements, similar to those of external doors. These are weather protection (including thermal insulation), security, ease of operation, durability and sound insulation. The extent to which these apply and their order of importance will vary from job to job. However, in every case the following points should be borne in mind. They can be considered the principles of window design, and apply mainly to timber windows:

- Use only durable, preservative-treated timber with suitable moisture content.

- Ensure all framing joints are tight fitting and use a waterproof adhesive.

- Prime or seal the whole assembly, preferably directly after manufacture.

- External horizontal surfaces should be weathered.

- The horizontal underside of projecting sills and transoms must be provided with a drip edge.

- The opening joint must be a tight fit for draught proofing, but must open out within rebate to avoid capillarity.

- Anti-capillary grooves and throatings should be incorporated around opening parts.

■ External glazing in painted work is best secured with weathered face putty and sprigs.

■ External glazing in clear finished work and internal glazing should be bedded in a suitable material but secured with cupped and screwed glazing beads from the inside of the building.

■ Rectangular window openings should have well-balanced proportions related to their height and width.

The majority of timber windows in use are made in a variety of standard styles and sizes by joinery manufacturers using volume production techniques to produce a standard product of known quality at an economical price. A typical example from a timber window manufacturer's range is illustrated in Figure 7.2. Rather than the mass production, the carpenter and joiner will be more involved with window construction from the following points of view: the like-for-like replacement of rotting windows in maintenance work; the one off or limited batch of windows for an extension to match those existing; the production of limited batch runs to a purpose-made design or size.

Window type		Frame dimensions width x height	
	V	634 x 921 634 x 1073 634 x 1226 634 x 1378	
	V	921 x 921 921 x 1073 921 x 1226	
	V	1221 x 1073 1221 x 1226	Side hung
	CV AS CV OPP	1221 x 1073 1221 x 1226	
	C AS C OPP	634 x 1073	
	CD AS CD OPP	1221 x 1073	
	CVC	1808 x 1073 1808 x 1226	Fanlight opening

Notes:
V = Ventlight   C = Casement   OPP = Opposite hand
D = Fixed light   AS = Hand as shown

▲ **Figure 7.2** Example from window manufacturer's range of casements

Casement windows and fixed lights

Casement windows are comprised of two main parts:

- *The frame,* which consists of head, sill and two jambs. Where the frame is subdivided, the intermediate vertical members are called mullions and the intermediate horizontal member is called a transom.

- *The opening casement,* which consists of top rail, bottom rail and two stiles. Where the casement is subdivided, both the intermediate vertical and horizontal members are called glazing bars.

Opening casements above the transom are known as **fanlights**.

Fixed glazing is called a dead light and glazing at the bottom of a window, normally below a casement is a sub-light. Where glass is bedded in the mainframe itself, it is known as direct glazing or direct lights.

The elevation of a four-light casement window is illustrated in Figure 7.3 with all of its component parts named. The 'four' refers to the number of glazed openings or lights in the window.

Casement windows can further be divided into two types, traditional and storm proof depending on their design and methods used in construction.

◀ **Figure 7.3** Four-light casement window

Traditional casements

These are fully rebated into and sit flush with the main frame. A vertical section through a traditional casement window is illustrated in Figure 7.4. Anti-capillary grooves are incorporated into the frame and

10 mm x 60 mm mortar key

70 mm x 95 mm head

45 mm x 45 mm top rail

10 mm x 15 mm glazing rebates

70 mm x 125 mm transom

6 mm anti-capillary grooves and drips

45 mm x 70 mm bottom rail

45 mm x 45 mm

45 mm x 70 mm bottom rail

70 mm x 145 mm sill

25 mm windowboard

▶**Figure 7.4** Traditional casement window (vertical section)

the opening casements in order to prevent the passage of water into the building. Drip grooves are made towards the front edges of the transom and sill to stop the water running back underneath them.

A mortar key groove is run on the outside face of the head, sill and jambs. The sill also has a plough groove for the internal windowsill or board to tongue into. Both the transom and sill incorporates a throat to check the penetration of wind-assisted rain. In addition this feature may be continued up the jambs. Finally the front of the transom and sill is weathered: it has a 9° slope for the rainwater to run off.

Figure 7.5 shows part of a horizontal section through a traditional casement window. It also shows the sizes and positions of the rebates, grooves and moulding in the jambs, mullion and the casement stiles.

All joints used in traditional casement window construction are mortise and tenons. Standard haunched mortise and tenons are generally used

70 mm x 90 mm mullion

70 mm x 95 mm jamb

45 mm x 45 mm stiles

▶**Figure 7.5** Traditional casement window (part horizontal section)

for the actual casements, although a sash haunch is preferable where smaller sections are used, see Figure 7.6. As a matter of good practice, the depth of the rebates should be kept the same as the depth of the mouldings. This simplifies the jointing, as the shoulders of the tenons will be level.

Figure 7.7 shows the jointing of head, jamb and sill of the main frame. These joints are normally wedged, although the use of draw pins or star dowels are acceptable and even preferable where the horn is to be later cut off.

Figure 7.8 shows an exploded isometric view of the joint between the transom and jamb. In order to make a better weatherproof joint, the front edge of the transom is housed across the jamb.

▲ **Figure 7.6** Haunched mortise and tenon joint secured with wedges

▲ **Figure 7.8** Transom joint details

Stormproof casements

These are partially rebated into and sit proud of the main frame. Storm proof casement windows incorporate two rebates, one round the main frame, and the other round the casement. These rebates, in conjunction with the drip edge, anti-capillary grooves and throat make this type far more weatherproof than traditional casements see Figure 7.9. Also shown is an alternative sill detail, which is used in conjunction with a stone, concrete or tiled sub-sill, when the window is set well back from the face of the wall.

Figure 7.10 shows types of seal, which can be fitted around the rebate of opening sashes to further improve the window's draft proofing qualities.

The jointing of the main frame of the stormproof casement window is often the same as that of the traditional casement, except for the transom, which is not housed across the face of the jambs since it is usually of the same width. Comb joints can also be used, although they leave no horn for building in.

Comb joints fixed with metal star dowels are normally used for jointing the actual casements (see Figure 7.11) although mortise and tenon joints can also be used.

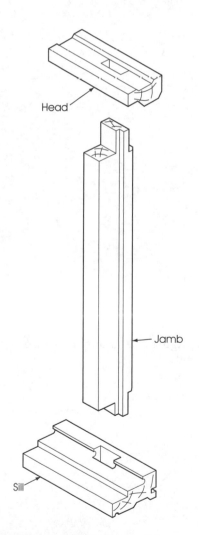

▲ **Figure 7.7** Main frame joints

Drip mould — 63 mm × 75 mm head
44 mm × 50 mm top rail
44 mm × 50 mm bottom rail
Drip mould
44 mm × 50 mm top rail — 63 mm × 75 mm transom
44 mm × 50 mm bottom rail — 63 mm × 75 mm sill
sill extension — 25 mm window board
44 mm × 50 mm bottom rail — Alternative sub-sill detail
63 mm × 75 mm sill
Terracotta tiled sill
63 mm × 75 mm mullion — 65 mm × 75 mm jamb
44 mm × 50 mm stile

▶**Figure 7.9** Stormproof casement window

Inward opening casements are rarely used, as they are very difficult to weatherproof. They also present problems for the occupier of the building because they obstruct the curtains.

Figure 7.12 shows a typical vertical section through an inward opening casement with a bottom hung inward opening fanlight. The bottom rail of the casement is rebated over a galvanised water bar, which is set in the sill. This prevents the passage of water into the building between the bottom rail and the sill. Inward opening fanlights are normally bottom hung and termed as 'hospital lights'.

Aluminium strip

Compressible rubber seal

Compressible rubber seal

Seal in aluminium carrier
surface fixed to inside of jamb

Seal strip ploughed into
jamb rebate

▲ **Figure 7.10** Draft seals for casement window

Top rail

Two tongues on rails

Comb joint

Three tongues on stiles

Metal star dowel

Stile

▲ **Figure 7.11** Comb joint secure with a
metal star dowel

inward opening fanlight

inward opening casement

►**Figure 7.12**
Inward-opening
casement

Timber sub-frames for metal casements

Metal casements can be fixed directly into the opening
in the wall. However, as these windows are produced
from light section material, they are better fixed in
timber sub-frames.

The timber frames are manufactured using mortise and
tenons or comb joints. Double rebates are incorporated
around the frame to provide a suitable bedding and
weather protection. A vertical and horizontal section of
a metal casement in a timber sub-frame is shown in
Figure 7.13. The metal casement is bedded in mastic
and screwed through the holes provided into the rebate
of the timber frame.

►**Figure 7.13** Metal casement in a timber sub-frame

head

metal
window

transom

sill

jamb

mullion

vertical section

horizontal section

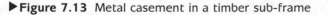

Combination frames

These combine in one frame a door and window. They are often used as entrance frames. The one shown in Figure 7.14 includes an outward opening external door, fixed glazing and side hung casement. Typical vertical and horizontal sections of a combination frame are illustrated in Figure 7.15. Where an inward opening door is required the rebates for this will have to be reversed and a water bar ploughed into the sill. Figure 7.16 shows an alternative to the solid middle rail. This is built up from two sections with a plywood panel ploughed between them.

▲ **Figure 7.14** Combination frame (door/window)

▲ **Figure 7.15** Combination frame details

French doors

French casements are often referred to as French doors. Figure 7.17 shows a French casement window with fixed sidelights and double opening casements. The construction of the frame is similar to that of the traditional casement frame or an external doorframe with a threshold. The opening casements and the fixed sidelights are constructed in a similar way to external glazed doors.

▲ **Figure 7.16** Alternative build-up middle rail

Robated meeting stiles

Mullion

Head

Jamb

Fixed side light

Double side hung casements (doors)

Threshold or sill

◀ **Figure 7.17** French casements

Bay windows

Bay windows are normally incorporated into a building for one or a combination of three reasons:

- to increase the amount of daylight and ventilation admitted into a room;
- to increase room size;
- to provide an architectural feature for the building.

The majority of bay windows are constructed using outward opening casements. The main construction details vary little from the standard casement windows covered above.

Bay windows are classified according to their shape on plan and are illustrated in Figure 7.18:

- square or rectangular bay
- cant or splayed bay
- segmental bay or bow window
- combination bay, square and cant.

Square bay

Cant bay

Segmental bay

Combination bay

▲ **Figure 7.18** Types of bay window

The angled intersections of the head and sill are normally mitred. These mitres are drawn up tight using handrail bolts. Two hardwood dowels are also used for each joint to overcome any tendency that the joint has to twist (see Figure 7.19).

The mullions or angle posts can be purpose made from solid timber, although for ease of manufacture these are often made in two sections. Alternatively, where standard windows are used to form a bay, separate corner posts can be used to join the jambs at the intersections (see Figure 7.20).

▲ **Figure 7.19** Joining head and sill

Two-piece (angle-post) mullion at intersection

Window jamb   Solid corner post

▲ **Figure 7.20** Mullion intersections

▲ **Figure 7.21** Building in a bay window

The preferred method of 'building in' the bay at the wall junction, is to keep the wall jambs square and angle the brickwork off to meet the frame, see Figure 7.21. Alternatively the back of the jamb may be built up and splayed off to meet the face brickwork.

A template should be provided with a bay window as an aid to the bricklayer when building the supporting wall. This should be made to the required brickwork sizes and not those of the actual window. Templates may be either cut from a sheet of plywood or made up using timber battens reinforced at the corners with plywood gussets as illustrated in Figure 7.22. The method used will depend on the material available and the window size.

▲ **Figure 7.22** Bay window template

Bow window

Segmental bay windows are often mistakenly called bow windows, but they are really a series of flat sections joined together to give a curved effect. Figure 7.23 is an illustration of a true bow window, often called a Georgian bow. Shaped cantilevered brackets may be required for support where the brickwork does not follow the window curve.

Typical sections for a direct glazed bow window are shown in Figure 7.24. The head and sill may be both cut and jointed from a solid section or vertically glulaminated to the required curve. Jambs and vertical glazing bars are mortised and tenoned into the head and sill. The horizontal glazing bars can be jointed to the jambs and vertical glazing bars using short stub tenons.

▲ **Figure 7.23** Bow window

▼ **Figure 7.24** Bow window sections

glulam head

horizontal glazing bars

glulam sill

vertical

jamb

vertical glazing bar

straight glazing rebate

part horizontal

Where the horizontal bars have a large curvature, short graining will weaken the tenons. In such cases they may be simply housed into the vertical members, or alternatively they may be also vertically glulaminated. The glazing rebate in the head, sill and horizontal bars is worked flat to avoid the necessity of using expensive special-order curved glass. In cheaper-quality work often only the head and sill are curved leaving the horizontal bars straight.

Roofs over bay and bow windows

These may be pitched and covered with tiles or slates to match the main roof or flat and covered with built up felt or sheet metal.

▲ **Figure 7.25** Roofs over bay windows

The construction of the roofs follows that covered in Chapter 2 of this book. Figure 7.25 illustrates a range of bay window roof configurations and details. Pitched roofs are all lean-to and may have hipped or gable ends. Where the roof is extended passed the window, possibly to form a porch over a front entrance, a timber gallows bracket may be required for support. Also available are one-piece GRP (glass fibre reinforced plastic) flat roofs, made to suit standard range of bay windows.

Curved-head casement windows

Where required, these can be constructed using the same method as that used for curved-head doors and frames. Curved members are built up from a number of pieces to avoid short grain, and jointed together with handrail bolts, false tenons, hammer-headed key and tenons or bridle joints. A typical semicircular-headed window with direct glazing is illustrated in Figure 7.26.

Glazing bars are used to subdivide the area into seven lights. These bars are cross-halved and scribed together where they intersect, then stub tenoned to the main frame. The radiating bars are stub tenoned at one end to a semicircular boss that is cut over and dowel-jointed to the cross-glazing bar.

▲ **Figure 7.26** Direct glazed semicircular-headed window showing section through boss

Pivot windows

Centre-hung pivot windows

These are often used for taller buildings, as both sides of the glass can be cleaned from the inside of the building with ease. The main disadvantage of this type is that, when opened, the top of the sash interferes with the curtains.

Figure 7.27 shows the vertical and horizontal sections of a traditional pivot window. Both the surrounding frame and the pivoting sash are constructed using mortises and tenons. The sash is hung on pivot pins about 25 mm above its centre line height to give it a self-closing tendency. The pivot pin usually is fixed to the frame and the socket to the sash. The planted stops, which form the rebates, serve to weatherproof the window. Those above the pivots are nailed or screwed to the frame on the outside and the sash on the inside, whereas those below are fixed to the frame on the inside and the sash on the outside.

45 mm × 45 mm
top rail

45 mm × 95 mm
head

12 mm × 19 mm
bead

45 mm × 95 mm
jamb

45 mm × 45 mm
stile

Groove to
enable sash
removel

45 mm × 70 mm
bottom rail

45 mm × 95 mm
sill

Vertical

Horizontal

▲ **Figure 7.27** Traditional pivot window sections

The actual positions of intersection of the beads needs to be precisely determined, especially where the sash is required to be removable without taking off any beads (for details of the method, see Book 1). The sash stile and its planted top bead must be grooved as shown to allow removal of the sash. The head and top rail is splayed to give sufficient opening clearance. This is provided at the sill by its weathering.

Stormproof pivot window

Figure 7.28 shows details of a stormproof centre-hung pivot window, which is an improvement on the type described above. The joints used in the frame are mortise and tenons. Comb joints are used for the sash. Face fixing friction pivots or back flap hinges are used for hanging the sash.

The moulded stop, which is mitred around the frame and cut on the pivot line, is glued and pinned to the top half of the frame and the bottom half of the sash.

Top part of sash open inwards

Moulded stop fixed to frame

Moulded stop fixed to sash

Lower part of sash and stop open outwards

▲ **Figure 7.28** Storm proof pivot window

Special windows

Proprietary windows

Specialist window manufacturers use various other proprietary designs of pivot windows. Figure 7.29 illustrates a typical section of such a window. Wide joints are used between the frame and the sash

▲ **Figure 7.29** Proprietary pivot window sections

throughout to combat capillarity. This is sealed with double-lipped
neoprene strips. External glazing beads have been used but at least
these are weathered and normally of hardwood. Hanging is by special
friction pivots, housed into the edge of both the frame and sash, the
knuckle of which is normally visible on the inside of the building.

Vertically hung pivot windows may be constructed using similar
details to those of the horizontal type. Although as the bottom pivot
takes all of the sash weight a special pivot is required which normally
incorporates some form of ball race mechanism.

Bull's eye windows

These are circular windows consisting of a main frame, which can be
glazed, direct or contain a sash. The sash itself may be either fixed or
pivot hung.

Figure 7.30 shows an elevation and section of a bull's eye window
with a pivot-hung sash that has been subdivided with glazing bars into
four lights.

There are various ways (already described in the section on curved
work with doors) in which the frame and sash can be made. In this case
they have both been formed from four curved segments, joined with
handrail bolts and dowels on the main frame with false tenons used for
the sash. The planted beads, which are fixed partly to the frame and
partly to the sash, have been glulamed using a number of thin
laminations bent around a suitable former. Both the frame and sash will
require flattening around the pivot positions on the inside to enable it
to pivot without binding. Threaded centre screw pivots are normally
used for bull's eyes, as these allow easy removal of the sash.

▲ **Figure 7.30** Bull's eye window (pivot hung)

▲ **Figure 7.31** Bull's eye window section (direct glazed)

Figure 7.31 shows a section of a direct glazed bull's eye window. In this case it has been built up in two layers with end joints staggered. The joint between the layers, which has been positioned to coincide with the rebate, is glued, screwed and pelleted. The glulamed glazing bead run around the inside of the frame is secured with recessed cups and screws.

Window ventilation

Ventilation to rooms is a requirement of the Building Regulations. Opening lights in the window normally provide rapid ventilation equal to 1/20th of the floor area, at least part of which must be at least 1.75 mm above floor level. Background ventilation of 8000 mm² for habitable rooms and 4000 mm² for kitchens and bathrooms is also required and again this must be at least 1.75 mm above floor level.

Vents

Background ventilation is normally provided by trickle type permanent vents incorporated into the window by one of the means shown in Figure 7.32:

■ in the head of the main frame;
■ in the top rail of the casement or sash;
■ within the glazed area itself.

Ventilators

These are used where permanent ventilation only is required and not daylight or a view. They consist of a main frame into which louvre slats are housed. The slats, which are normally set at 45°, should project beyond the face of the frame to throw rainwater clear. The top and

240

Air flow

Slotted ventilator in head of frame

Opening light to provide rapid ventilation equal to 1/20 of floor area

Slotted ventilator in top rail of casement

Air flow

Air flow — Adjustable ventilator in glazed area — Control cord

◀ **Figure 7.32** Window vents

bottom edges of adjacent slats should overlap to prevent through vision and restrict the passage of driving rain (see Figure 7.33).

The elevations and sections of the two most common ventilators are illustrated in Figure 7.34. The triangular frame may be jointed using dovetails or pinned bridles and the circular one built up using any of the methods covered previously. The true shapes of the louvre slats and the housings in the frame must be determined geometrically (see Book 1). This is because the slats are inclined and do not intersect with the frame at right angles.

Assembly procedure for window frames and opening sashes

This can be carried out for all the above types of frames and opening sashes using the following assembly procedure, see Figure 7.35:

1. Dry assemble to check fit of joints, overall sizes, square and winding.

2. Clean up inside edges of all outer framing components and both edges of inner ones.

3. Glue, assemble, cramp up and wedge. Re-check for square and winding.

frame

louvre slat

overlap

45°

projection

▲ **Figure 7.33** Typical ventilator (vertical section)

▲ **Figure 7.34** Louvre ventilators

triangular   section

circular   section

▲ **Figure 7.35** Assembling casement sash

4. Clean up remainder of item and prepare for finishing.
5. Hang, where used, opening casement sashes and fanlights.

Hanging casement sashes

Storm proof casements fit on the face of the frame and normally require no fitting at all. The only operation necessary to hang the casement is the screwing on of the hinges. Traditional casements fit inside the frame and require both fitting and hanging, see Figure 7.36.

▲ **Figure 7.36** Hanging of casements sashes

▲ **Figure 7.37** Fitting and hanging a casement sash

The hanging of casement sashes follows a similar procedure to that used for doors and is illustrated in Figure 7.37

1.   Mark the hanging side on both the sash and the frame.

2.   Cut off the horns.

3.   'Shoot in' (plane to fit) the hanging stile.

4.   Shoot the sash to width.

5.   Shoot in the top and bottom of the sash.

6.   Mark out and cut in the hinges.

7.   Screw one leaf of the hinges to the sash.

8.   Offer up the sash to the opening and screw the other leaf to the frame.

9.   Adjust fit if required and fix any other ironmongery.

Sliding Sash Windows

These may be classified into two groups according to their direction of opening:

- vertical sliding sash window
- horizontal sliding sash window.

Vertical sliding sashes

These consist of two sashes, which slide up and down in a main frame. They are also known as double-hung sliding sash windows. There are two different forms of construction for these types of window:

- those with boxed frames
- those with solid frames.

Boxed frames

This type of window is the traditional pattern of sliding sashes and for many years has been superseded by casements and solid frame sash windows. This was mainly due to the high manufacturing and assembly costs of the large number of component parts. An understanding of their construction and operation is essential, as they will frequently be met in renovation and maintenance work.

The double-hung boxed window consists of two sliding sashes suspended on cords or chains which run over pulleys and are attached to counterbalanced weights inside the boxed frame.

Figure 7.38 shows an elevation and horizontal and vertical sections of a boxed frame sliding sash window. It shows the make-up of this type of window and names the component parts.

Inside elevation   Horizontal section   Vertical section

▲ **Figure 7.38** Boxed frame sliding sash window details

Head lining

Slot for wagtail

Stile housed and nailed into head

Pulley stile wedged into sill

Wagtail hangs from head lining

Sill cut away for outside and inside lining

◀ **Figure 7.39** Boxed frame joint details

The pulley stiles are housed into the head and sill. These are simply skew nailed at the head and wedged at the sill as illustrated in Figure 7.39. The sill will require cutting away on both ends, to accommodate the linings. Wagtails which separate the weights to prevent them fouling in use are simply pinned into slots cut in the head.

The inside and outside linings, which are butt jointed at the corners, are tongued and nailed to the head and pulley stiles. Glue blocks are used to strengthen the joints at the head, but must not be used in the stiles, as they will interfere with the sash weights, see Figure 7.40. The outer linings, which extend past the pulley stiles and head, often have a bevelled edge, requiring a mitre to be incorporated in the butt joint.

- Figure 7.41 illustrates how the outside lining and parting bead at sill level are cut away in order to prevent water and dirt being trapped at this point.
- Figure 7.42 shows an alternative detail, which can be used at sill level; it uses a deeper draughtboard tongued into the sill replacing the standard staffbead. This allows the bottom sash to be partly opened to provide ventilation at the meeting rails without causing a draught.
- A pocket piece is cut in the pulley stiles to provide access to the weights. It is cut out by making saw cuts as shown in Figure 7.43 and then firmly tapping out from the back of the stile.

Glue block to reinforce joint

Linings butted

Linings nailed to head and stile

Mitre required when edge of lining is moulded

▲ **Figure 7.40** Boxed frame head details

parting bead

draught board

outside lining and bead cut away to prevent water lodgement

alternative sill

▲ **Figure 7.41** Cutting away of outside lining and parting

▲ **Figure 7.42** Alternative sill detail

pulley stile

saw cuts

pocketpiece

saw cuts

screw

▲ **Figure 7.43** Pocket piece

■ Figure 7.44 shows the mortise and tenon joint used between the sash stile and rail. The sash haunch or reverse franking is used instead of a normal haunch. If this were not done, the joint would be seriously weakened by the sash cord groove that runs into the edge of the stile.

■ Figure 7.45 shows two alternative methods used for the joints between the stiles and meeting rails. Where the horn is left on it is usually moulded and a mortise and tenon joint can be used. A dovetail joint should be used where no horn is required.

▲ **Figure 7.44** Joint detail (sash haunch)

▲ **Figure 7.45** Alternative meeting rail joint details

▲ **Figure 7.46** Boxed double mullion (plan view)

Where a window is more than one sash wide, boxed double mullions shown in Figure 7.46 are introduced to accommodate the four weights. These have a wide bulky appearance, so they were often concealed on the outside with ornamental stone mullions or pilasters.

Venetian windows

As shown in Figure 7.47, Venetian windows do not require boxed mullions as normally only the middle sashes slide, with the narrow side sashes being screw fixed permanently in the closed position.

▲ **Figure 7.47** Venetian window (inside elevation)

▲ **Figure 7.48** Solid mullion section and method used to hang centre sashes

- Figure 7.48 shows a section through the solid mullion of a Venetian window and the method used to hang the centre sashes. The pulleys are fixed to the pulley stiles as high as possible. The cords fixed at one end to the sashes pass over the mullion pulleys, across the heads of the side sashes, then finally over the side pulleys to the weights.

- Figure 7.49 shows how the cords are concealed across the top. The top rail of the side sash is grooved out to receive the outer cord. A grooved cover mould that also incorporates the top staff bead conceals the inner cord. Both the cover mould and the fixed top sash must be made removable for cord replacement.

▲ **Figure 7.49** Concealing sash cords across head

Curved head boxed frame sashes

The inside elevation of a semicircular headed vertical sliding sash window is shown in Figure 7.50. The construction details of this sash window are the same as those of a standard sash up to the springing line. Above this a number of alternative methods may be used to construct the frame and sash.

Figure 7.51 shows the curved head built up in three layers with joints well staggered and glued together. The middle layer, often of plywood, also forms a fixed parting bead. Below the springing line a removable parting bead is used. Also illustrated is how the sash stile to curved rail joint (handrail bolt with dowels) is positioned beyond the springing to enable a shoulder to be formed. The shoulder stops on the projecting edge of the built-up head when the sash is lifted. This prevents jamming and impact at the crown, thus avoiding the risk of sash damage and breaking glass.

An alternative method of forming the curved head is shown in Figure 7.52. In this case the head has been glulamed from thin solid laminates or thin plywood. Blocks are glued around the head to provide a fixing for the curved inner and outer linings, which, unlike the straight ones, are merely butt jointed and pinned, not tongued and grooved. The parting bead in this case would be shaped from ply or solid timber and grooved into the head.

▲ **Figure 7.50** Semicircular-head sash (inside elevation)

247

▲ **Figure 7.51** Semicircular-head sash details

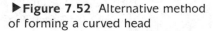

►**Figure 7.52** Alternative method of forming a curved head

Balancing of sashes

For all forms of boxed sash windows, two cast iron or lead weights are required to balance each sash. To determine the size of these weights the fully glazed sashes can be weighed with a spring balance. The weights for the top sash should each be half the sash weight plus about 250 g and the bottom sash weights half the sash weight less 250 g. Using this principle, both sashes are counterbalanced, but will have the tendency to remain closed in their respective positions, firmly against the head or sill.

Solid frame sash windows

Sash balances are manufactured for use in place of sash cords and weights, which do away with the need for boxed frames, thereby, reducing the number of component parts and simplifying the construction of sliding sash windows.

Figure 7.53 illustrates the elevation, horizontal and vertical sections of a solid frame sliding sash window. It can be seen from this that apart from the solid jambs, the arrangement of the component parts is similar to that of the boxed frame sash window. The frame and sashes can be jointed using mortise and tenon joints, although modern mass-produced windows are now exclusively manufactured using comb joints with metal star dowels. For lightweight domestic sashes, the spring balances are accommodated in grooves run in the back of the sash stiles. The

Head
Staff bead
Spring balance
Outside lining
Top rail
Meeting rail
Bottom rail
Draught board
Sill
Jamb
Staff bead
Parting bead

Top screw to jamb
Spring balance
Bottom screwed to underside of stile and rail
Spring balance grooved into frame jamb
Alternative for heavy sashes
Stile
Spring balance grooved into sash stile

▲ **Figure 7.53** Sash window sections

spring balances for heavyweight industrial sashes are accommodated in grooves run in the actual jamb of the frame.

Two balances are required for each sash. Where the sashes are the same size, the longer balances are for the bottom sash and the shorter balances are for the top. The balances are fixed to the tops of the jambs with a screw. The plate on the other end is attached to the bottom rail of the sash. As each sash is raised or lowered the spring is tensioned. The sash is then supported or balanced by the tension of the springs.

After glazing the sashes, the springs should be adjusted to balance the sash and ensure a smooth operation. If the sash is not balanced both springs on the sash should be adjusted. Unscrewing the sash plate and twisting it one turn in a clockwise or anticlockwise direction will adjust the tension of the springs. The plate should then be re-screwed to the sash and the operation of the sash checked to see if it is correct.

Horizontal sliding sashes

The elevation and sectional details of a window with horizontally sliding sashes is shown in Figure 7.54. The main frame is simply constructed from rectangular section material using mortises and tenons. It has planted staff beads and grooved parting beads. The sashes, which are also framed using mortises and tenons, have rebated meeting stiles to prevent draughts. Both sashes may be made to slide on a track or waxed hardwood runners.

▲ **Figure 7.54** Horizontal sliding sash window

Larger three or four sash windows can be constructed, using similar details to those shown. Normally only the middle sash or sashes slide, while the outer ones are fixed to the main frame.

Assembly procedure for boxed frames

The sashes are normally made up first as they can be used when checking the boxed frame for size. After preparing and cleaning up all of the components and fitting of the pulleys, the following assembly procedure can be used for the boxed frame, see Figure 7.55:

1. Assemble basic frame: glue and wedge pulley stiles to sill; glue and nail head to pulley stiles; place on level bench bearers and check for square and wind.

2. Glue and pin the outer linings to the stiles and head. Fix glue blocks to reinforce the head joints.

3. Turn the frame over and repeat Step 2, this time for the inner linings.

4. Check sashes for size and shoot to fit if required, so that they slide easily. A working clearance of 2 mm on both sides should be sufficient to allow for subsequent finishing. Remove sashes for glazing and weighing.

5. Mitre the parting beads and dry position, insert the head before the sides.

6. Mitre the staffbeads and dry nail in position, leaving the nail heads protruding.

▲ **Figure 7.55** Assembly of a boxed frame

The window may be supplied to site in either as a boxed frame for building in, with the sashes separate for site hanging, or as a fully glazed and hung unit. In both situations the cording of the sashes can be undertaken using the procedure covered later in Chapter 12 of this book.

Glazing to windows

Windows are single or double-glazed using clear, or patterned (obscure) glass. Glazing to windows below 800 mm from the floor level should use a safety glass that will not produce pointed shards with razor shape edges if broken. It is also recommended that environmental control glass be used. In temperate climates with cool winters such as in the UK where thermal insulation is most important, **low emissivity (Low E) glass** is used. This has a coating that allows heat from the sun to enter the building but reduces the amount of heat loss the other way from inside the building. Used mainly as the inner pane of double glazed windows with the coated side facing into the gap or air space. In warmer climates a **solar control glass** may be used to limit the amount of heat gain from the sun into the building.

Single glazing has been traditionally used for windows, 4 mm or 6 mm thickness glass being the norm, which is held in position with either putty or glazing beads, see Figure 7.56. The procedure to be used is covered later in Chapter 12 of this book.

▶**Figure 7.56** Single glazing

Double glazing

This is now preferred as it offers higher levels of both thermal and sound insulation. In addition, as the temperature of the inner pane is warmer, the risk of condensation on the room side is reduced. This can be achieved using either sealed double-glazing units, or secondary glazing.

Sealed double glazing units

These are often used to double glaze windows for thermal insulation purposes. These consist of two panels of glass with a gap of up to 25 mm between them. The panes are held apart by a spacer that runs around the perimeter and is sealed to provide an airtight joint. The units are assembled in a clean dry atmosphere or an inert gas is used to fill the gap to avoid any possibility of condensation forming between the panes, see Figure 7.57. In addition the greater mass of sealed units compared to single glazing, will reduce to some extent the passage of sound.

Wider rebates are normally required around the edges of casements and sashes to accommodate sealed units. However where the rebate in existing windows is too small to take the extra thickness of glass, stepped units can be used. Figure 7.58 shows how normal and stepped sealed glazed units are bedded and held in position. Bedding can be done using either self-adhesive strips or a non-setting glazing compound. Spacer blocks are required along the bottom of the rebate to support the unit and at the back of the rebate all around the perimeter when using a glazing compound to prevent it squeezing out as the unit is pushed home. Beads are the preferred method of securing the units, either screwed or nailed as appropriate.

▲ **Figure 7.58** Fixing sealed double glazing units

Figure 7.59 illustrates a method of double-glazing that is similar in effect to sealed units. This is achieved by fixing a second pane of glass on the inside of the sash or casement with the aid of rebated beads. A distinct disadvantage of this method is the very high risk of condensation occurring between the two panes of glass.

Secondary double-glazing

Whilst sealed units do provide a certain amount of sound insulation, this is often insufficient for noisy locations, for example, near busy roads or airports. In general, the sound insulation properties of a window increase as the air space between the two windows increases.

▲ **Figure 7.59** Alternative to sealed double glazing units

▶**Figure 7.60** Secondary double glazing

For efficient sound insulation, the air space should be between 150 mm and 200 mm. In order to achieve this air space, a second window can be built on the inside of the reveal behind the existing window. This type of double-glazing is known as secondary double-glazing.

Figure 7.60 shows a secondary double-glazing detail. As a further improvement, the reveals between the two frames can be lined with a sound absorbent material, such as strips of insulation board.

Linked casement windows

These provide an alternative method of double glazing which is popular in Europe (see Figure 7.61). The two casements are hinged and screwed together. The outer one is normally hinged to the frame. Where the casements are screwed together, the fixings should be accessible, as the inner faces of the glass will

vertical section

part horizontal section

◀**Figure 7.61** Linked casements

require cleaning occasionally. This can be reduced to a minimum if the two casements are sealed together with an air permeable but dust excluding sealing strip.

Condensation can occur between the two panels of glass where this space is not ventilated. However, ventilation has the effect of reducing the window's insulating value.

Window ironmongery

The selection of window fittings will depend on its type and location. Figure 7.62 illustrates a range of commonly used items that are readily available in a variety of materials and finishes.

- *Butt hinges* are used for traditional side hung casements.
- *Cranked hinge.* The cranked hinge is also known as the *storm-proof hinge* as it is used for hanging rebated storm proof casements.
- *Offset hinge.* These can be used for hanging both traditional and storm-proof casement windows. When open they allow an arm to be passed through the gap between the frame and the casement in order to clean the outside of the window from the inside, hence the popular name 'easy clean' hinge.
- *Window pivot.* These pivots are recessed into the edge of the window jamb and stile. They are used for the concealed hanging of pivot windows.
- *Friction pivot.* This is a type of hinge used for hanging pivot windows. It screws on to the face of the window and also contains a friction action, allowing the window to remain open in any position.
- *Side opening mechanism.* These pivot and slide the sash from the side, creating a gap at the jamb for ease of cleaning the outside. Versions are available that enable the sash to tilt as well as turn.
- *Casement stay.* This is also known as a *peg stay*. It is used to hold the window in various open positions.
- *Casement fastener.* This is also known as a *cock spur* and is used to secure a casement window in the closed position. Versions are also available with a key-operated lock for additional security.
- *Window lock.* A key-operated lock fitted to casements and fanlights for additional protection.
- *Fanlight catch.* Fitted to the top rail of horizontal pivots and inward opening fanlights to secure them in the closed position. For high-level windows the ring pull may be operated with a window pole or cord system.
- *Quadrant stay.* Used in pairs to limit travel of inward opening fanlights.
- *Sash fasteners.* These are available in a number of types and are used to hold the meeting rails of a sash window together, thereby securing the sashes in a closed position. It is recommended that two are fitted to ground floor windows for additional security.

▲ **Figure 7.62** Window ironmongery

- *Retaining stay.* Used to limit the travel of horizontal pivots and fanlights.
- *Security screw.* Fitted mainly to the meeting rails of sliding sash windows for extra security. The threaded screw passes through the socket drilled into the inner meeting rail and locates in the plate fixed to the outer meeting rail. It can also be used to secure casement windows by passing through the casement stile and into the jamb.
- *Sash lift.* One or two are fixed to the bottom rail of a vertical sliding sash to act as a finger grip when lifting or lowering the sash.
- *Sash pulley.* Various patterns are available to take different sized cords and chains. They are mortised and recessed flush into the top of the pulley stile.

Fixing windows

In common with doorframes, widows may be fixed in position by either building in during the main building process or 'fixed in' at a later stage. The actual method used will depend on the window construction, the material used (hardwood and plastic windows are usually fixed in), and the availability of the windows at the wall construction stage.

Fixing casement window frames

Built in

The bricklayer fixes the majority of timber frames by building in as the brickwork proceeds, as shown in Figure 7.63.

1. The frame sill is bedded level using bricklayer's mortar and the head is checked for level and packed up as required.

2. Temporary struts are used to hold the frame upright.

3. Jambs should be plumbed from the face. It is standard practice to plumb and strut the first using a spirit level. The other is then sighted parallel: stand to the side of the frame, close one eye, sight the edge of the plumbed jamb with the edge of the other and adjust if required until both jambs are parallel.

4. As the brickwork proceeds, galvanised metal frame cramps should be screwed to the back of the jambs and built into the brickwork. Three or four cramps should be evenly spaced up each jamb.

5. The horns of mortise and tenoned frames should be cut back for 'building in', rather than cut off flush, which help to fix the frame in position. The horns of comb-jointed frames are often cut off flush with the jambs. After trimming the horns it is essential that the cut ends be treated with paint or preservative in order to reduce the possibility of timber rot.

▲ **Figure 7.63** Built in frames

Fixed in

Where window frames are fixed in at a later stage the carpenter is often required to make a timber template or profile frame. These are made up either on-site or in the workshop using 50 mm × 75 mm softwood and plywood corner gussets as shown in Figure 7.64.

Temporary struts to hold profile plumb

Vertical 'DPC' separates inner and outer walls

Plywood gusset

Inner block work

Softwood frame

Cavity

Outside brickwork

Packer for tolerance and ease of removal

▶ **Figure 7.64** Profile frame for windows that are to be fixed at a later stage

The profile is made about 6 mm wider than the actual window to give a fixing tolerance. The vertical tolerance is achieved by setting the profile on 6-mm packing blocks to aid removal. After the wall has set the profiles may be removed, leaving the prepared opening ready to receive the frame.

1. The frame is placed in the prepared opening by temporarily holding it in position with the aid of folding wedges (Figure 7.65). The actual position of the frame in the reveal will depend on the type of frame and presence of a sub-sill. Standard casement windows are fixed with about a 25 mm setback from the face of the wall; those with flush sills that are intended to be used in conjunction with a stone, concrete or tiled sub-sill are set back about 110 mm.

2. Check the sill for level and the jambs for plumb; adjust the wedges as required.

3. Fix the frame to the wall reveal by the use of normal plugs and screws or proprietary sleeved frame fixing plugs and screws. As an alternative fixing for timber windows, metal plates may be prefixed to the back of the jambs before positioning the window and subsequently secured to the wall by plugs and screws. A minimum of three fixings per jamb will be required, whichever method is used.

4. Seal the gap around the frame with a silicone or acrylic frame sealant. These are normally supplied in cartridges and gun applied. The seal should be forced as far back into the gap as possible, finally finishing off with a narrow angled face bead. Any slight unevenness in the face bead can be smoothed using a small paintbrush dipped in water.

Figure 7.65 Fixing windows in prepared openings

Fixing sash windows

Solid frame sash windows are usually built in in the same manner as casements, using galvanised frame cramps. Boxed frame sash windows are normally fixed into a prepared opening in the wall by driving wedges in behind the jambs (see Figure 7.66).

▲ **Figure 7.66** Fixing boxed frame sash windows

Window boards

A window board is normally used to finish the top of the wall internally at sill level. They may be formed from solid timber, plywood, MDF or plastic and should be fitted at the first fixing stage before the internal walls have been plastered. Where the window board fits up to a timber window, it is normally tongued into the sill groove. If the window is metal or plastic (uPVC) the window board is only butted to the sill, with the gap being filled with caulking or covered with a small bead, see Figure 7.67.

Many window boards have a groove on the underside; this provides a finish for the plaster and masks any gap which may result later from shrinkage. Alternatively, a cover mould may be used to mask the joint.

▲ **Figure 7.67** Window board details

Cutting and fixing window boards

See Figure 7.68:

1. Cut a length of board about 100 mm longer than the sill.

2. Mark out each end and cut out a portion of the board to fit the window reveal.

3. Return the front edge nosing profile around the ends, using a block plane and finish with an abrasive paper.

4. Use an off-cut of window board and a boat level to check for front to back level. Packing may be required under the board where it is not level; proprietary plastic shims available in a range of thicknesses are the best. However a piece of hardboard or several pieces of damp-proof material will also do the job. These should be positioned about 50 mm from each end, with intermediates spaced at about 450 mm centres

5. Place the window board in position and fix the front edge to the wall, using either cut nails, masonry nails or plugs, screws and pellets. Fixings should preferably go through any packers so the board is not pulled down out of line and also to ensure they are not misplaced later.

Waste to be removed

Distance between window reveals

Depth of reveal

Check front to back level

Returned nosing profile

Cut nail fixings through packings

Packing as required

Window board nailed to ground

Ground nailed to top on wall

Frame anchors screwed to underside of window board and face of wall

'No nails' type adhesive used to bond pre-finished window boards

◄ **Figure 7.68** Fixing window boards

Alternative fixing methods

A batten or ground may be pre-fixed to the wall at the correct level for the window board, using cut or masonry nails, the window board in turn being fixed directly to the ground using ovals nails. Plastic and other pre-finished window boards may be bedded and fixed to the wall using beads of a gap-filling 'no nails' type adhesive, or frame anchors screw-fixed to the underside of the window board and face of the wall.

Chapter Eight

Mouldings and trim

This chapter covers the work of the site carpenter and joiner. It is concerned with the recognition and installation of mouldings and trim. It includes the following:

- common types
- cutting trim
- installation of architraves for door openings
- installation of skirting
- installation of dado and picture rails.

Mouldings are the ornamental contours or shapes applied to angles or features of building material for decorative purposes. They are variously named according to their profile, see Figure 8.1.

Trim is the collective term for timber moulded sections. They may be termed as either vertical or horizontal and are used to cover the joint between adjacent surfaces, such as wall and floor/ceiling or the joint between plaster and timberwork. In addition they can provide a decorative feature, and may also serve to protect the wall surface from knocks and scrapes as illustrated in Figure 8.2.

Timber trim and architrave types

The common range and location of timber trim is illustrated in Figure 8.3.

- *Architrave*. The decorative trim that is placed internally around door and window openings to mask the joint between wall and timber and conceal any subsequent shrinkage and expansion.

- *Plinth block*. The block of timber traditionally fixed at the base of an architrave to take the knocks and abrasions at floor level. Also used to ease the fixing problems at this point that can occur when the skirting is thicker than the architrave.

- *Skirting*. The horizontal trim that is fixed around the base of a wall to mask the joint between the wall and floor. It also protects the plaster surface from knocks at floor level.

- *Dado rail*. A decorative moulding applied to the lower part of interior walls at about waist height approximately 1 m from the floor. It is also known as a chair rail as it coincides with a tall chair

▲ **Figure 8.1** Mouldings

◀ **Figure 8.2** Covering of gaps and protecting plasterwork

back height to protect the plaster. It also provides a line for the application of texture wallpaper down to the skirting, which was a feature in Victorian and Edwardian times.

- *Picture rail.* A moulding applied to the upper part of interior walls between 1.8 m and 2.1 m from the floor. Special clips may be hooked over the rail in order to suspend picture frames.

- *Cornice.* The moulding used internally at the wall/ceiling junction. It is normally formed from plaster, only rarely from timber. The area between the cornice and the picture rail is known as the *frieze*.

Cutting and fixing trim

Solid timber, either softwood or hardwood, is used for the majority of mouldings. However, MDF (medium density fibreboard) and low density foamed-core plastics are used to a limited extent for moulding production. They are normally ready to install, machined to a range of standard profiles (shapes) by the manufacturer or supplier as illustrated in Figure 8.4.

Skirting is often mass-produced using a combination profile, for example, with an ovolo mould on one face and edge and a splayed and rounded mould on the other. This enables it to be used for either purpose and also reduces the timber merchant's stock range. In addition, the moulding profile has the effect of an undercut edge enabling it to fit snugly to the floor/wall junction.

The following cutting and fixing details are generally suitable for all three of these materials. Consult manufacturer's instructions prior to starting to fix other proprietary mouldings/trim.

▶ **Figure 8.4** Standard trim sections

Installing architraves

Architraves are normally cut and fixed after hanging the doors. A set of architraves for one side of a room door, consists of a horizontal head and two vertical jambs or legs, see Figure 8.5.

A 6 to 9 mm **margin** is normally left between the frame or lining edge and the architrave (Figure 8.6). This margin provides a neat appearance to an opening; an unsightly joint line would result if architraves were to be kept flush with the edge of the opening.

Mitres

The return corners of a set of architraves are mitred, see Figure 8.7. For right-angled returns (90°) the mitre will be 45° (half the total angle) and can be cut using a mitre box or block. Mitres for corners other than right angles will be half the angle of intersection. They can be practically found by marking the outline of the intersecting trim on the frame/lining or wall, and joining the inside and outside corners to give the mitre line (see Figure 8.8). Mouldings can be marked directly from this or alternatively an adjustable bevel can be set up for use.

Fixing mitres. Mitres should be nailed through their top edge to reinforce the joint and ensure both faces are kept flush (Figure 8.9). 38-mm oval or lost-head nails are suitable for this purpose.

▲ **Figure 8.5** Architraves

▲ **Figure 8.6** Margin to architraves

◄ **Figure 8.7** Mitring architraves

▲ **Figure 8.8** Determining mitre for corners other than right angles

Mark margin

Cut first mitre, mark second

Cut second mitre and fix head

Mark mitre to first jamb

Cut mitre, fix first jamb

Mark mitre to second jamb

Cut mitre, fix second jamb

Punch in nails and remove arrises

▲ **Figure 8.11** Marking and fixing architraves

▲ **Figure 8.9** Nailing mitre joints at corners of architraves

Cutting and fixing architraves to door openings

Fixing is normally direct to the door frame/lining at between 200 mm and 300 mm centres using typically 38 mm or 50 mm long oval or lost-head nails. These should be positioned in the fillets or quirks (flat surface or groove in moulding, see Figure 8.10) and punched in. In addition, very wide architraves are often fixed back to the wall surface using either cut or masonry nails.

▲ **Figure 8.10** Fixing architraves

Do not forget to wear eye protection when driving masonry nails.

A typical fixing procedure is as follows:

- Mark a margin line around the door opening (Figure 8.11). Check-lines the width of the architrave may also be marked on the wall at the top of the opening.

- Starting with the headpiece, cut the mitre on one end, hold the piece in position and mark the second mitre.

- Cut the second mitre and temporarily fix in position with two nails. Leave the nails protruding in case later adjustment is required.

- Hold the first jamb in position and mark the mitre. Cut the mitre, fit with a block plane if required and temporarily nail in position.

- Repeat the last stage with the second jamb.

- Complete nailing at 200 mm to 300 mm intervals, including nailing across the mitres. Punch all nails below the surface in preparation for filling later by the painter.

- Use a glass paper block to remove the arris around the frame, architraves and mitres.

Plinth blocks. Where used, these may be butt jointed to the architrave, as shown in Figure 8.12 but traditionally they were pre-joined together using a barefaced tenon and screws.

Corner blocks. Sometimes used at the return corners of a set of architraves in place of mitres (Figure 8.13). These may be simply pinned in place and the architrave butt jointed to them. Alternately they may be prefixed to the architrave using dowels or a biscuit. To mask the effect of subsequent block shrinkage the head and leg architraves may be housed into them.

▲ **Figure 8.12** Plinth block

▲ **Figure 8.13** Use of corner blocks

▲ **Figure 8.14** Scribing architraves

Scribing architraves

Architraves should be scribed (one member cut to fit the contour of another) to fit the wall surface, where frames or linings abut a wall at right angles.

- Temporarily fix the architrave jamb in position, keeping the overhang the same all the way down, see Figure 8.14.

- Set a compass to the required margin plus the overhang, or alternatively use a piece of timber this size as a gauge block.

- Mark with the compass or gauge up the line to be cut. Saw along the marked line by slightly undercutting the edge (making it less than 90°); it will fit snugly to the wall contour, see Figure 8.15.

▲ **Figure 8.15** Scribing should fit snugly to wall surface

Quadrant and Scotia moulds

A quadrant mould or Scotia mould is often used to cover the joint to provide a neat finish to the reveal of external door frames (Figure 8.16). These moulds may also be used in place of an architrave jamb where the frame/lining joins to a wall at right angles. The arris on the back of the mould is best pared off with a chisel, to enable the mould to sit snugly into the plaster/timber intersection.

All nails used for fixings moulds should be punched below the surface on completion of the work. This is in preparation for subsequent filling by the painter.

Sharp arris on back of quadrant removed, fits snugly to wall

▲ **Figure 8.16** Use of quadrants as an alternative to scribing

Installing skirting

Skirting is normally cut and fixed directly after cutting and fixing the architraves.

Jointing skirting

Internal corners

Corners that are less than 90° are best scribed, one piece being cut to fit over the other, see Figures 8.17 and 8.18.

▲ **Figure 8.17** Internal corners

▲ **Figure 8.18** Scribing internal corners

Scribes can be formed in one of two ways:

■ *Mitre and scribe.* Fix one piece and cut an internal mitre on the other piece to bring out the profile (Figure 8.19). Cut the profile square on mitre line to remove waste. Use a coping saw for the curve.

■ *Compass scribe.* Fix one piece and place the other piece in position (Figure 8.20). Scribe with a compass. Cut square on the scribed line to remove waste.

Scribing is the preferred method especially where the walls are slightly out of plumb. The mitred scribe would have a gap, but the compass scribe would fit the profile neatly.

Mitres are not normally used for internal corners of skirting, as wall corners are rarely perfectly square, making the fitting difficult. In addition, mitres open up as a result of shrinkage, forming a much larger gap than scribes.

45°

Mitre end
in box

Skirting profile

Cut square on profile line

◀ **Figure 8.19** Cutting an integral
scribe (mitre and scribe)

Fixed piece of skirting

Mark profile
with compass

Cut square on profile line

Neat fit

Gap

Compass scribe

Mitre scribe

▲ **Figure 8.20** Compass marking and cutting a scribe

However, internal corners on bull-nosed or pencil-rounded skirtings
may be cut with a **false mitre** or partial mitre on the top rounded
edge and the remaining flat surface scribed to fit (Figure 8.21).

Internal corners over 90°. These are called obtuse angles
(see Figure 8.17) and are best jointed with a mitre.

Mitring external corners

These return the moulding profile at a corner (Figure 8.22), rather than
a butt joint which would otherwise show unsightly end grain. Mitres for
90° external corners may be cut in a mitre box or with the aid of a
frame saw or mitre chop saw.

Mitres for both internal and external corners over 90° can be marked
out using the following method and then cut freehand:

False
mitre

Scribe

▲ **Figure 8.21** False mitre and
scribe

▲ **Figure 8.22** Mitring external corners

- Use a piece of skirting to mark line and width of skirting on floor, either side of the mitre (Figure 8.23).
- Place length of skirting in position.
- Mark position of plaster arris on top edge of skirting. For internal corners this will be the actual back edge of the skirting (Figure 8.24).
- Mark outer section on front face of skirting.

▲ **Figure 8.23** Line of skirting for corners over 90°

▲ **Figure 8.24** Marking out external and internal corners over 90°

- Use try square to mark line across face and back surface of skirting.
- Cut mitre freehand or with the aid of a frame saw or chopsaw set to the required angle.
- Repeat marking out and cutting procedure for other piece.
- Fix skirting to wall; external corners should be nailed through the mitre.

Long lengths are fixed first (Figure 8.25), starting with those having two trapped ends (both ends between walls). Marking and jointing internal corners is much easier when one end is free.

Where the second piece to be fixed also has two trapped ends, a piece slightly longer than the actual length required, by say 50 mm, can be angled across the room or allowed to run through the door opening for scribing the internal joint (Figure 8.26).

▲ **Figure 8.25** Order of fixing (trapped pieces first)

◀ **Figure 8.26** Extend through doorway to permit scribing of joint, or angle and scribe when both ends are trapped

After scribing and cutting to length it can be fixed in position.

Very short lengths of skirting returned around projections may be fixed before the main lengths. The two short returns are mitred at their external ends, cut square at their internal ends and fixed in position (Figure 8.27).

◀ **Figure 8.27** Small pieces may be fitted first

Main wall lengths are then scribed and fixed in position. These help to hold the short returns. Finally the front piece is cut and fixed in position by nailing through the mitres.

Heading joints can be used where sufficiently long lengths of skirting are not available. Mitres are preferred to butts, because the two surfaces are held flush together by nailing through the mitre (Figure 8.28). In addition mitres mask any gap appearing as a result of shrinkage.

Butted heading joint
gaps on shrinkage
and hard to keep flush

Mitred heading
joint flush and
gaps masked

▶ **Figure 8.28** Mitres are preferred for heading joints

Installing skirting on a curved surface

Where skirtings and other trim are to be fixed around a curved surface, the back face will almost certainly require **kerfing**. This involves putting saw cuts in the back face at regular intervals to effectively reduce the thickness of the trim. The kerfs, which may be cut with a tenon saw, should be spaced between 25 mm and 50 mm apart. The tighter the curve, the closer together the kerfs should be. The depth of the kerfs must be kept the same. They should extend through the section to the maximum extent, but just keeping back from the face and top edge (Figure 8.29).

Care is required when fixing trim back to the wall to ensure that it is bent gradually and evenly. It is at this time that accuracy in cutting the kerfs evenly spaced to a constant depth is rewarded. Any over-cutting and the trim is likely to snap at that point.

Back of
curved
section
kerfed

Depth
of kerf

25 mm to 50 mm

Kerfs
evenly
spaced

▶ **Figure 8.29** Fixing to a curved surface using kerfing

Mitres at the ends of curved sections may be marked out and cut using the same methods as described for other obtuse angles.

Returned stop mouldings

Skirtings and other mouldings may occasionally be required to stop part way along a wall, rather than finish into a corner or another moulding such as skirting to a landing. In these circumstances the profile should be returned to the wall or floor.

■ *Returned to wall.* This can be achieved by mitring the end and inserting a short mitred return piece giving a true mitre for good quality work. Alternatively, the return profile can be cut across the end grain of the main piece. However, end grain will remain exposed (Figure 8.30).

▲ **Figure 8.30** Return the profile of moulding that stops part way along a wall

■ *Returned to floor.* This involves mitring the moulded section (Figure 8.31), cutting out the waste and inserting a return moulded section (cut from an off-cut) down the end of the board.

Fixing skirtings

Skirting can be fixed back to walls with the aid of:

■ grounds

■ timber twisted plugs

■ direct to the plaster/wall surface.

Grounds

These are timber battens which are fixed to the wall surface using either cut nails in mortar joints or masonry nails. One ground is required for skirtings up to 100 mm in depth. Deeper skirtings require either the addition of vertical soldier grounds at 400 mm to 600 mm centres or an extra horizontal ground (Figure 8.32). The top ground should be fixed about 10 mm below the top edge of the skirting.

Packing pieces may be required behind the grounds to provide a true surface on which to fix the skirting. Check the line of ground with either a straightedge or string line. Skirtings can be fixed back to grounds using, typically, 38-mm or 50-mm oval or lost-head nails.

▲ **Figure 8.31** Stopped moulding returned to floor

Single
ground
for small
skirting

Packing as required

Double
grounds
for taller
skirting

Alternativly
a single
ground and
soldiers

Horizontal
ground

Soldier
ground

400 to
600mm centres

▶ **Figure 8.32** Timber grounds for fixing skirting

Plugs marked ready to be cut

Thickness of plaster

Shape of timber plug

▲ **Figure 8.33** Fixing skirting to timber plugs

Twisted timber plugs

These are rarely used today. They are shaped to tighten when driven into the raked-out vertical brickwork joints at approximately 600 mm apart (Figure 8.33). When all the plugs have been fixed, they should be marked with a straightedge or string line and cut off to provide a true line. An allowance should be made for the thickness of the plaster.

Skirtings are fixed back into the end grain of the plugs using, typically, 50-mm cut nails. These hold better in the end grain than would oval or lost-head nails.

Direct to the wall

Skirtings are fixed back to the wall after plastering, using typically either 50-mm cut nails or 50-mm masonry nails depending on the hardness of the wall. Oval or lost-head nails may be used to fix skirtings to timber studwork partitions.

Fixings should be spaced at 400-mm to 600-mm centres (Figure 8.34). These should be double nailed near the top and bottom edge of the skirting or alternatively they may be staggered between the top and bottom edge. Remember all nails should be punched below the surface.

Hardwood skirtings for very high quality work may be screwed in position. These should either be counter-bored and filled with

Fixings at
400 mm to 600 mm
centres

◀ **Figure 8.34** Spacing of fixings

cross-grained pellets on completion, or brass screws and cups should
be used (Figure 8.35).

As a modern alternative to nails and screws, skirting and other trim
fixed to wall surfaces may be bonded using a gun-applied gap filling 'no
nails' adhesive. One or two continuous 6-mm beads should be applied
to the back of the trim before positioning and pressing in place.
Strutting or temporary nails (Figure 8.36) may be required to hold the
trim in place whilst the adhesive cures. These should be left overnight
before removal.

Cup and
screw

Counter-bored
and pelleted

▲ **Figure 8.35** Screws sometimes
used to fix hardwood skirtings

Temporary
strutting

'No nails'
gap-filling
adhesive

Temporary
nail

◀ **Figure 8.36** Use of gap-filling
adhesive to fix skirting

Checking for hidden/buried services

Prior to fixing mouldings across any wall, a check should be made to
see if any services are hidden below the wall surface. Fixing into electric
cables and gas or water pipes is potentially dangerous and expensive to
repair. Wires to power points normally run vertically up from the floor.
Wires to light switches normally run vertically down from the ceiling.
Therefore keep clear of these areas when fixing. Buried pipes in walls
are harder to spot.

Vertical pipes may just be seen at floor level; outlet points may also be
visible (Figure 8.37). Assume both of these run the full height of the
wall, both up and down. Therefore, again keep clear of these areas
when fixing. When in doubt an electronic device can be used to scan
the wall surface prior to fixing. This gives off a loud noise when passing
over buried pipes and electric cables.

▲ **Figure 8.37** Locating hidden services

▲ **Figure 8.38** Keeping skirting tight to floor surface when fixing

Holding down and scribing skirtings

When fixing narrow skirtings, say 75 mm to 100 mm in depth, they may be kept tight down against a fairly level floor surface with the aid of a kneeler (Figure 8.38). This is a short piece of board placed on the top edge of the skirting and held firmly by kneeling on it.

Deeper skirtings and/or uneven floor surfaces may require scribing to close the gaps before fixing. This is carried out if required after jointing but prior to fixing:

- Place cut length of skirting in position.
- Use gauge slip or compass set to widest gap to mark on the skirting a line parallel to the uneven surface (Figure 8.39).
- Trim skirting to line using either a handsaw or plane.
- Undercut the back edge to ensure the front edge snugly fits the floor contour.

Gauge slip equal to widest gap under skirting

Edge undercut for snug fit to floor

▲ **Figure 8.39** Scribing and cutting skirting to an uneven floor surface

Installing dado and picture rails

Being mainly horizontal mouldings, they may be cut and fixed using similar methods to those used for skirting.

Start by marking a level line in the required position around the walls (Figure 8.40). Use a straightedge and spirit level or a water level and chalk line. The required position may be related to a datum line where established.

▲ **Figure 8.40** Marking position of dado and picture rails

When working single-handed, temporary nails can be used at intervals to provide support prior to fixing (Figure 8.41).

Simple sections can be scribed at internal corners, as are skirtings; otherwise, use mitres. External corners should always be mitred.

Fixings are normally direct to the wall surface at about 400 mm centres using typically either 50 mm cut nails or 50 mm masonry nails depending on the hardness of the wall. 50 mm oval or lost-heads can be used when fixing mouldings to timber studwork partitions. Alternatively a gun-applied gap filling 'no nails' adhesive may be used to bond them to the wall surface.

Mitres around external corners should be secured with nails through their edge. Typically 38 mm ovals or lost heads are used for this purpose. Remember, the carpenter should punch all nails below the surface ready for subsequent filling by the painter.

▲ **Figure 8.41** Temporary support for dado and picture rails during fixing

Raking moulds

These are mouldings that are not fixed horizontally, such as those that follow the pitch of the stairs. Where raking moulds join up with a level moulding at a mitred corner, the actual profile of one mould will need to change. Either a special mould is required or more practically the level mould can be returned around the corner a short distance and mitred to the raking mould of the same section as shown in Figure 8.42. See Book 1 for raking mould geometry.

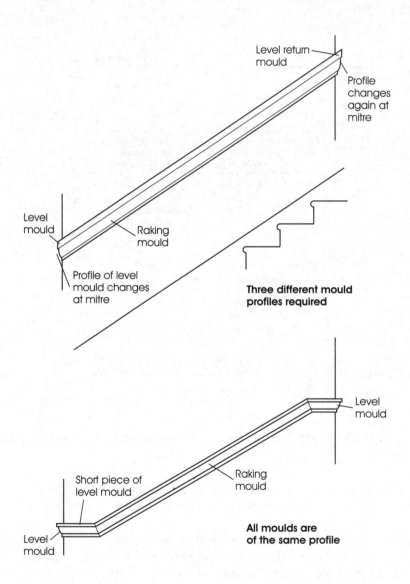

▲ **Figure 8.42** Raking moulds

Chapter Nine

Units, fitments, seating and miscellaneous items

This chapter covers the work of both the site carpenter and bench joiner. It is concerned with principles of constructing and installing units, fitments, seating and other items. It includes the following:

- ❡ types, manufacture and installation of purpose-made and proprietary units
- ❡ types, manufacture and installation of built-in and free-standing fitments
- ❡ retail units and fitments
- ❡ types and manufacture of seating units
- ❡ miscellaneous joinery items.

Timber units and fitments here comprise cupboards, built-in or fixed-in with worktops, shelves, drawers and doors and special-purpose joinery items including public seating.

Timber units

Units fall into two distinct categories:

- ■ *Purpose made.* A unit made in a joiners shop for a specific job. Most will be fully assembled prior to their arrival on site.
- ■ *Proprietary.* A unit or range of units mass-produced to standard designs by a manufacturer. Budget-priced units are often sent in knockdown form (known as *flat packs*) ready for on-site assembly. Better quality units are often ready assembled in the factory (known as *rigid units*).

Terminology

- ■ *Cupboard.* Historically derived from the name given to the simple open boards or shelves which were used to display cups, silver plate and other items ('cup board').
- ■ *Carcass.* The main assembly or frames of a cupboard, excluding doors and drawers.
- ■ *Pot board.* The lower shelf or base of a cupboard. Again historically derived from the name given to a low board or shelf, raised just off the floor on which heavy cooking pots were stored.

- *Plinth.* The recessed base of a cupboard, which supports the pot board. It also provides a foot space for those standing in front of the cupboard and is often called a kickboard.
- *Standards.* The vertical end frames or panels and intermediate divisions of a cupboard. In some parts end standards are known as 'gables' and intermediate standard as 'haffits'.

Methods of construction

The two main methods of carcass construction for both proprietary and purpose-made units, shown in Figure 9.1, these being:

- box construction
- framed construction.

▶Figure 9.1
Types of unit construction

Box construction

Framed construction

Box construction

This is also known as slab construction. It uses vertical standards and rails and horizontal shelves. A back holds the unit square and rigid. The plinth and potboard are often integral with the unit.

- *Proprietary units* are almost exclusively made from 15-mm to 19-mm thick melamine-faced chipboard (MFC) or medium density fibre board (MDF).
- *Purpose-made units* may be constructed using MFC, MDF, blockboard, plywood or, more rarely, solid timber.
- *Flat packs* use **knock-down fittings** or screws to join the panels. Assembly is a simple process of following the manufacturer's instructions and drawings, coupled with the ability to use a screwdriver. See Figure 9.2.

Rigid and purpose-made units may be either doweled or housed and screwed together. Glue is used on assembly to form a rigid carcass. Typical carcass construction details are illustrated in Figure 9.3.

Framed construction

This is also known as **skeleton construction**. This uses frames either front and back joined by rails or standards, or cross frames joined by rails. The plinth and potboard are normally separate items.

Sleeve

Connecting screw to secure units together

Recessed support fitting

Pin

Cam

Trim cap

Pin

Pin and can panel fixing. Use trim cap when seen

One or two piece plastic knock down blocks

▲ **Figure 9.2** Box construction unit using typical knock-down fittings

The frames of proprietary units are normally dowelled, whereas purpose-made ones would be mortised and tenoned together. Typical carcass construction details are illustrated in Figure 9.4.

Adjustable shelves

In situations where the sizes of the items to be stored are not known or where they are subject to change, some form of shelf adjustment must be incorporated. Illustrated in Figure 9.5 are three of the many methods used:

- The traditional solution was the use of saw tooth supports and splayed-end push-in shelf bearers that can be fitted at any desired height.

- A very popular and efficient method is the use of **tonks**: bookcase strips and studs, which allow height adjustment in 25-mm units. The flush strip is designed to be recessed into the standards. A smaller, deeper groove must also be run to give clearance for

Back pinned
to standard

Back grooved
into end standard

Rail screwed to
intermediate standard

Overhang allows
for scribing and
can accomodate
services

Rail dowelled
to end standard

Standard housed
and screwed to
potboard

Potboard

End standard
housed or
dowelled to
potboard

Glue
blocks

Separate plinth
overlaid with
laminate after
screwing together

▶**Figure 9.3** Rigid box construction

inserting the tongues of the studs. Alternatively, a surface-fixed strip
can be used. This overcomes the need for grooving out and
weakening members, but results in an inferior finish.

■ Sockets tapped into blind holes, which have been drilled at intervals
down the standards, and used with push-in studs, are suitable for a
lighter range of applications.

Drawers

Drawers may be incorporated into units in order to provide storage and
security (when fitted with a lock) for smaller items. The size of a
drawer will be related to the items it is intended to store, but in general

Back grooved or rebated

Framing joined using mortise & tenons, stubbed into panel groove or dowels

In fill to end frames: Grooved, rebated or overlaid

Framed end standard

Cross rails dovetailed tenoned or dowelled to end standard

Cross rails screwed or tenoned to intermediate standard

Plinth screwed to end standard and overlayed or combined with separate potboard

▲ **Figure 9.4** Framed unit construction details

will range between 100 mm and 200 mm in depth. When they are vertically stacked in one unit the deeper drawers should be located at the bottom.

Figure 9.6 illustrates a traditional method of drawer construction, which uses through dovetails at the back and lapped dovetails at the

▶**Figure 9.5** Adjustable shelf details

Splayed end bearer

Saw tooth support

Tonks strip

Stud

Stud

Socket

Blind holes

▼ **Figure 9.6** Drawer construction

Hand-cut dovetails

Machine-cut dovetails

front. The plywood bottom is grooved into the front and sides and is pinned to the bottom of the drawer back. Small glue blocks are positioned under the bottom to provided additional rigidity and assist sliding. Rounded machine-made dovetails, produced on a router or spindle moulder, are a more economic alternative for better quality work where repetitive production is required.

Modern mass-produced drawers are made in MFC or MDF. The corner joints can be housed and pinned, butted and screwed, dowelled or biscuited. The bottom of the drawer is typically either nailed or screwed directly to the underside of the drawer sides. A separate false slab front is secured to the drawer by screws from the inside as illustrated in Figure 9.7.

Dowels

Biscuits

Butted
and
screwed

Bottom
pinned or
screwed

Housed
and pinned

False slab
front screwed
through drawer

▲ **Figure 9.7** Modern drawer construction

Various methods can be used to suspend and slide drawers. Figure 9.8 shows how traditionally rails may be incorporated into a unit for this purpose. The dust board shown grooved into the rails is mainly used on better quality work. Its purpose is to separate the drawer and cupboard spaces. A drawer kicker is fitted between the top rails to prevent the front of the drawer falling downward as it is pulled out. When closed, the drawer front should finish flush with the unit. This can be achieved by pinning small plywood drawer stops to the front of the drawer rail.

Drawer kicker

Dust board

Drawer rail

Drawer stop

▶**Figure 9.8** Drawer rail details

Another traditional method of suspending and sliding drawers is shown in Figure 9.9. This uses grooved drawer sides, preferably of a hard wood with good wearing qualities, which slide on hardwood runners glued and screwed to the unit's sides or standards. Where this method is used, the drawer is often fitted with a false slab front screwed from the inside of the drawer. This front has projecting ends to conceal the runner from view. Alternatively, fibre drawer slides may be used with one part fixed to the side of the drawer and the other to the unit.

Modern methods of suspending and sliding drawers, shown in Figure 9.10, employ the use of metal, side or bottom mounted runners. These incorporate a ball race or plastic roller to ensure ease of operation.

Grooved
drawer
side

Hardwood
runner

False
front

Fibre
slide

▲ **Figure 9.9** Drawer slide details

Bottom mounted

Side mounted

▲ **Figure 9.10** Metal drawer runners

Doors

Doors can be incorporated to close the front of open units for reasons of tidiness, protection or security. They may be either side-hung or sliding doors. Side-hung doors allow maximum access but when they are open, they project into the room, which can be restrictive and even hazardous in confined spaces. Figure 9.11 illustrates various methods of side-hanging cupboard doors:

- *Type A: flush hinge.* The door is set flush within the unit and hung on butt hinges. The use of flush hinges avoids the need for recessing and provides the necessary clearance joint.

▶ **Figure 9.11** Side-hung cupboard doors

(A) Flush hinge

(B) Cranked Hinge

(C) Pivot hinge

(D) Modern concealed hinge

Vertical adjustment

Sideways adjustment

Front to back adjustment

■ *Type B: cranked hinge.* Rebated doors hung on cranked hinges were at one time popular for mass produced units. These doors do not require any individual fitting as the rebate, which laps over the face of the unit, conceals the very large clearance joint.

■ *Type C: pivot hinge.* Doors hung on the face of a unit are probably the simplest to make and fit, although this arrangement is possible with standard butt hinges, the use of cranked or special extended pivot hinges permits the door to open within the width of the unit.

■ *Type D: concealed hinge.* Doors are again face hung. A concealed cabinet hinge is bored into the rear face of the door and its mounting plate fixed to the inside of the standard. This type, which is extensively used in modern kitchen units, allows the door to be adjusted in height, sideways, backwards and forward. They include a degree of self-closing in the final stage of swing and so do not require a catch.

Sliding doors

There are various methods available for making cupboard doors slide. Glass, thin plywood or MDF doors are often made to slide in nylon or fibre tracks (see Figure 9.12). The deeper channel track is used at the top, so the doors can be inserted and removed by pushing them up into the top track, clearing the bottom one. Figure 9.12 also shows how a cupboard door may be made to slide on fibre tracks grooved into the potboard. Two nylon sliders are recessed and screwed to the underside of each door. These run on the fibre track and provide a smooth sliding action, which wears well. The top edge of the door is usually rebated to engage in a groove run in the underside of the top rail. Sufficient clearance for insertion and removal must be made at this point to allow the bottom of the door to clear the potboard when pushed up into the

◀ **Figure 9.12**
Glass and wooden
sliding cupboard doors

Fibre
track

Alternative
top guide
(retractable)

Fibre
track

Nylon slider
(two to a door)

top groove. Retractable top guides similar to flush bolts can be used instead of rebating the top edges of the doors.

Heavyweight cupboard doors are best top hung to achieve a smooth running action. Figure 9.13 illustrates one of the simplest types of top-hung cupboard door sliding track. It consists of a surface-fixed aluminium top track and bottom guide. The door is suspended by two nylon hangers/sliders fixed to its top edge. The handle position of bottom-sliding doors is best kept nearer the bottom of the door for smooth sliding action. There will be a tendency for the door to tip and judder if the handle is positioned higher than a distance equal to the door's width. The best action is achieved with top-hung doors when the handle position is kept as high as possible.

Worktops

The main types that are in common use are illustrated in Figure 9.14.

- *Solid timber.* Made up from narrow-edge jointed boards. Provision for moisture movement needs to be considered when fixing them to a carcass. Mitred return ends may be specified; these serve to hold the top flat and also cover the exposed end grain.

- *Post-formed.* A chipboard base covered with a plastic laminate, which has been formed over a rolled edge. The most popular type of

▲ **Figure 9.13** Top-hung sliding door

▶ **Figure 9.14** Worktop details

worktop for kitchen units, it is ready finished and simply requires fixing in place.

- *Edged sheet material.* A chipboard or MDF base covered with either a melamine face, plastic laminate or wood veneer. Matching edging is applied to the seen edges. The undersides of sheet material tops should be the same as the top face or be sealed to prevent distortion. It is good practice to use a double-faced, melamine or wood veneer board. On good quality work a balancer laminate is applied to the underside of laminate-faced tops to relieve the stresses that would result from facing one side only. A cheaper alternative is to seal the underside with an application of varnish or adhesive.

- *Wood trimmed.* A chipboard or MDF base covered with either melamine face, plastic laminate or wood veneer. A hardwood lipping or edging is simply glued; rebated, glued and screwed; or tongued and glued to the seen edges, providing a neat finish. Return corners are better mitred than butted.

- Wood trimmed and over-laid. A chipboard or MDF base edged in hardwood and overlaid with either laminate or wood veneer. The laminate is trimmed off on the spindle moulder or by use of a hand-held power router. Often a decorative edge feature is incorporated in the trimming process.

Built-in fitments

These normally use the wall, floor and/or ceiling as part of the construction.

Cupboards

Figure 9.15 shows details of a cupboard that is built into a reveal at the side of a fireplace. It has been framed up using 38 mm × 75 mm framing. The skirting is continued across the front of the frame to match in with the existing timber work. The doors, which have rebated meeting stiles, are also framed up and a 9-mm plywood is used for the panels. The top, base, shelf and front framework are fixed to 25 mm × 50 mm battens that have been plugged and screwed to the walls.

Wardrobes

Shown in Figure 9.16 are details of a built-in wardrobe. This can be made up using 18-mm or 25-mm MFC or MDF for the base, partitions, shelves and doors. To provide a good finish, the exposed edges should be either lipped with 10-mm timber edging or taped with an iron-on edging. See Figure 9.17.

The outside doors are hung on 25 mm × 50 mm battens which have been fixed to the walls, while all the remaining doors are hung on the partitions. These can be hung inside the opening using flush hinges, or on the face using concealed cabinet hinges.

The base is made up on a plinth board to match the height of the existing skirting as shown. An in-fill piece is used at the top to drop the head of the wardrobe down, so that a cornice to match the existing one can be fixed along the ceiling line if required.

Block board or MDF
top and shelves

Moulded trim
to front of top

Bearers
fixed to
wall

Mortise &
tenoned
framed
doors

Mortise &
tenoned
front
frame

Skirting to match
remainder of room

Rebated
meeting
stiles

Plywood
panel

▲ **Figure 9.15** Framed front built-in fitment

Assembly procedure for units and fitments

The assembly of framed items follows closely that of other framed joinery items. The following procedure is illustrated in Figure 9.18:

1. Dry assemble to check fit of joints, overall sizes, square and winding.

2. Clean up inside edges of all framing components and both faces of infill panels, etc.

3. Glue, assemble, cramp up and wedge each individual frame. Re-check for square and winding.

4. Clean up internal faces of individual frames.

5. Clean up any rails, shelves, top and potboard, etc.

6. Glue, assemble and cramp up individual frames and other members to form the unit carcass. Check for square.

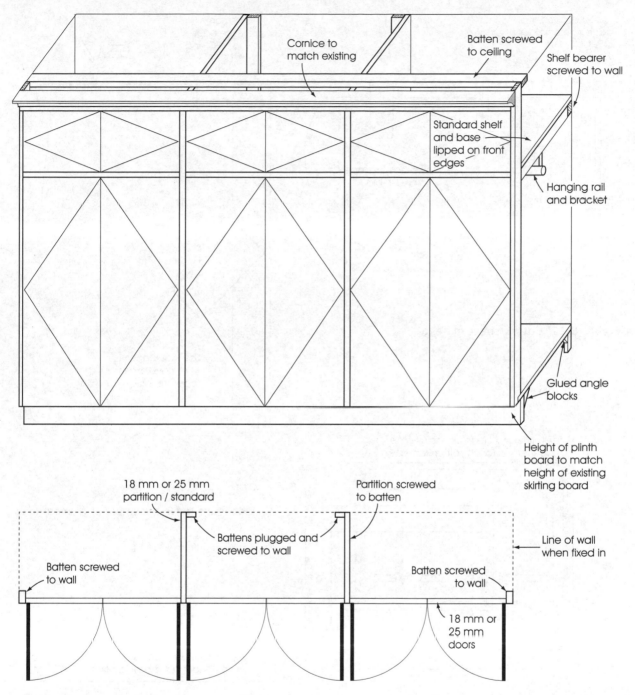

▲ **Figure 9.16** Built-in wardrobe

7. Clean up external surfaces and prepare for finishing.

8. Fix worktop if separate.

9. Install drawers, hang doors and fix any ironmongery.

The assembly of box construction units can be carried out using the following procedure shown in Figure 9.19.

1. Check measurements of panels.

2. Carry out any necessary handwork, such as squaring out corners, iron-on edging, fittings for shelves and drawers, etc.

Glued on timber
edge to blockboard
or MDF

Use masking tape
to hold in place
while glue dries

Iron-on
tape edge
to MFC and
MDF

Use glasspaper
to remove excess
tape and arris

▲ **Figure 9.17** Edging sheet material

Assemble and clean up main frames
using method for any other framed joinery

◄ **Figure 9.18** Assembly of framed units

Apply glue
to joints

Cramp up
if required

Assemble frames and other members
to form carcass

Fix work top, hang doors
fir ironmongery and insert shelves

▲ Figure 9.19 Assembly of box construction

3. Where proprietary knock-down fittings are being used, these should be pre-fitted to each panel at this stage.

4. Assemble panels, use glue and cramps if required.

5. Fix worktop if separate.

6. Install drawers, hang doors, fit any ironmongery and insert any shelving.

Retail units and fitments

Although there are many variations to suit a wide range of retail outlets, hotels, bars, office receptions, etc., they may all be constructed using a similar carcass. The main differences are size, front treatment and worktop or counter top treatment.

▶ **Figure 9.20** Pictorial view of bank/building society counter

Figure 9.20 is a pictorial impression of a bank or building society counter. Wide counter tops and glass screens are used as a security measure. They separate persons (i.e., staff and customers) and property, ensuring each remains on their respective side.

A section through a typical bank/building society counter is shown in Figure 9.21. Veneered blockboard lipped with a matching hardwood on exposed edges and housed together has been used for the main carcass. This sits on a separately framed plinth. The counter top, which is also made from blockboard, has a boxed-out overhanging front edge, finished with a deep hardwood edge trim. Leather, PVC, lino or plastic laminate could be used for the actual top finish. The top rails are continuous over the intermediate standards, which should be cut out to receive them. The front of the counter has been given a decorative feature by fixing hessian- or PVC-covered blockboard panels to the carcass. Alternatively, a more conservative, traditional effect can be achieved by fixing framed dado panelling to the front of the counter carcass. The staff side of the counter is fitted with drawers and sliding doors.

Figure 9.21 Section through bank/building society counter

Bar counters and fitments

A typical bar counter and over-bar unit is illustrated in Figure 9.22. This incorporates a lifting counter flap and inward

counter flap catch

lifting counter flap

counter flap hinge

bar door

▲ **Figure 9.22** Bar counter and over-bar unit

opening door to permit through access. When closed, the flap and door should be continuous with the front and top surface so that they are hardly noticeable. Counter flap hinges are used to hang the flap. These have no projection above the counter surface when the flap is closed and allow the flap to open through 180° to rest flat on the counter top when required. When the flap opens against a wall a counter flap catch should be fitted to hold it in the open position.

A section through the bar counter showing its construction is illustrated in Figure 9.23. The inside of this counter and any other used for food should have easily cleaned hygienic surfaces. In this case, all internal surfaces have been covered with plastic laminate including the faces of the plinth. Also indicated is how a washing up sink can be incorporated below the bar.

The bar top itself has been formed from a solid piece of 32-mm hardwood thickened at its front edge to give a deeper impression. This would be fixed to the main carcass using

solid timber bar top

lipped blockboard

provision for washing up sink

all internal surfaces covered with plastic laminate

▲ **Figure 9.23** Section through bar counter

▲ **Figure 9.24** Mitred counter end cramp

▲ **Figure 9.25** Counter cramp for heading joint

lipped
blockboard

all internal
surfaces covered
with plastic
laminate

corner
trim

brass
support pole

▲ **Figure 9.26** Over-bar unit

shrinkage plates and finished with polyurethane varnish or other suitable treatment to give a heat, water and spirit resistant surface.

Solid timber bar tops and other counter tops of any width will probably require a number of pieces to be edge-jointed together, normally by a well glued joint incorporating a loose tongue for strength and to keep the faces flush during assembly. The ends can be cramped as shown in Figure 9.24. This mitred cramp, into which the top is tenoned, serves to keep the top flat and prevents any unsightly end grain showing.

Heading joints in a worktop are sometimes required where long runs or a change of direction are necessary. These can be formed using a loose tongue or dowels to keep the top surface flush. Traditionally, the joint is pulled up tight with a counter cramp that remains permanently under the counter (see Figure 9.25). Today, worktop connection bolts are more common.

It can be seen that the cramp consists of three short pieces mortised to take wedges. They are screwed to the underside of the top across the joint initially at one end only, the two outside pieces to one half, and the middle piece with its mortise slightly off centre to the other half. After driving the wedges to pull up the joint, the remaining screws can be driven. Alternatively small blocks could be fixed to the underside of the top and a sash cramp then used to pull up the joint.

Construction details of the over-bar unit are shown in Figure 9.26. Again this utilises lipped veneered blockboard housed together and surfaced on the inside with plastic laminate. This unit is positioned above the bar and held by brass support poles. The ends are fixed back to the structure in order to provide it with a degree of stability.

Curved counters

Figure 9.27 shows how curved counters may be formed using curved rails, shelves, pot board and plinth. The ends and intermediate standards are positioned so that they radiate normal to the curve. For obvious reasons (short graining) both the rails and the plinth cannot be cut in the usual way from solid timber. Instead they may be made from plywood, MDF, or formed using glulam methods. The counter front could be finished in 6-mm bendy plywood or MDF bent around the curve. This would be glued and pinned to the main carcass. Alternatively the facing may be built up using a double layer of 3-mm ply or MDF on tight curves.

Screens

Dwarf screens or partitions, less than the full height of a room, are used in a variety of buildings to provide visual separation between adjacent areas. Figure 9.28 shows an elevation and sectional detail of a fixed dwarf partition about a metre in height. It consists of a framed panelled section fixed between two built-up end posts. A grooved capping for additional stiffening tops this. A skirting to match that of the room is fixed on both sides.

standards normal
to curve

glulam, MDF
or plywood
curved rails

battens

separate
plinth

▲ **Figure 9.27** Curved counter section

Where the partition is continuous between two walls, a door or **wicket gate** may be provided for access. The hanging and closing edges would be rebated to prevent a through joint and also act as an effective stop for the closed position. A radiused rule joint, as shown in Figure 9.29, will have to be used on the hanging edge of the overhanging cap to prevent binding on opening. Alternatively parliament hinges could be used, as these would extend the pivot point outside the capping.

Figure 9.30 illustrates details of a typical free-standing **movable screen** often used to divide up and give a degree of privacy to open-plan offices. The framework, which incorporates the feet, is of polished hardwood, while the in-fill panels fixed to softwood battens may be of veneered plywood (as shown) or plywood covered with PVC, hessian or similar material. Alternatively, the in-fill panels can be made of softboard or other acoustic material with the hollow core between them filled with fibreglass or rock wool. This would give them excellent sound absorption qualities, making them ideal for screening noisy machinery or other activities. In addition this type of screen is suitable as a pinboard for poster or information display purposes.

elevation

capping

planted mould

skirting

vertical section

built-up post

framing

panel

part horizontal section

▲ **Figure 9.28** Dwarf partition

rule joint in capping

rebated joint in partition and gate

line of capping

line of dwarf partition and gate

◀ **Figure 9.29** Wicket gate

elevation

softwood batten

hardwood frame

veneered plywood panel

typical section

▶ **Figure 9.30** Free-standing screen

Fixing joinery items

This category includes screens, partitions, cupboards, etc. These will have been made in a joiner's shop, and most are fully assembled prior to their arrival on site. However, budget-priced units (kitchen cupboards, wardrobes etc.) are often sent in flat pack form for simple on-site assembly, normally involving either screws and dowels or proprietary joint devices.

Flat pack unit assembly and installation

The method of assembling and installing flat-pack units will vary from manufacturer to manufacturer. However, each unit is supplied with its own instructions. It is most important to take the time to read through these prior to commencing work. In general this is a three-stage process:

■ *Assembly*. Put carcasses together. Unpack and assemble units one at a time and check contents. Open more than one and you risk confusing the parts!

■ *Installation*. Fix base units to the wall starting with a corner base unit and working outwards from either side. Finally install wall units, again working from the corner outwards.

■ *Finishing*. Fit worktops, drawers and doors. This should not be done until all units are firmly fixed to wall and connected together.

Fixing base units

See Figure 9.31:

▲ **Figure 9.31** Fixing base units

1. Position base units level and plumb using wedges provided if required. Ensure all top edges and fronts of units are flush. Secure units together using connecting screws. The best place to start is in the corner.

2. Drill, plug and screw units to wall, using the brackets provided.

3. Trim floor wedges flush to unit with a knife.

Fixing wall units

See Figure 9.32:

1. Draw a level horizontal line from the top of the tall unit if being used. It is normal practice to keep tall units and wall units level with each other. Draw another line the depth of the wall units below this line. This marks the position of the underside of the wall units. Where tall units are not being used, wall units are normally fixed with a gap of 450 mm between their underside and the work surface.

2. Temporarily fix a batten on the marked level line, to act as a support while marking fixing holes and screwing.

3. Rest the unit on the batten. Ensure the tops of wall units and tall units are flush.

4. Mark the wall through the fixing holes.

5. Drill, plug and screw the unit to the wall.

▲ **Figure 9.32** Fixing wall units

6. Packing behind a fixing may be required on an uneven wall surface to ensure the units are plumb.

7. Position the remaining wall units in place one at a time.

8. Ensure the top edges and fronts are flush and secure together using connecting screws.

Fixing worktops

See Figure 9.33 for post-formed worktops see 'Worktops', in the previous section for other types.

1. Measure and cut the worktop to the required size.

2. Metal filler/joint strips or routed butt and mitre joints are used to connect worktops in corners. These joints can be pulled up tight using worktop connecting bolts.

3. Position worktop and screw in place through the fixing brackets.

4. Sawn bare ends of the top can be covered with a metal trim or a plastic pre-glued edge banding. Iron the edge banding into place, using a sheet of paper in between banding and iron for protection.

◀ **Figure 9.33** Fixing post-formed worktops

Fixing drawers

Insert drawers, made up previously, onto drawer runners.

Fixing doors

1. Lay the doors face down on a flat clean surface.

2. Locate the hinges over the previously fixed hinge plate and secure hinges with a mounting screw.

3. Adjust the hinges if required to ensure accurate door alignment.

Rigid and purpose-made unit installation

The installation of rigid and purpose-made units follows the general procedures used for flat-pack units.

Good quality rigid units have adjustable legs for easier levelling on an uneven floor. In addition, wall units often have adjustable wall brackets that enable fine adjustment to plumb and level, see Figure 9.34.

Mitred at corners

Cut mitre from face in adapted mitre box

Metal hanging plate fixed to wall

Cornice where used screwed to top of units

Adjustment screws for levelling

Adjustable bracket fixed to top inside corners of wall unit

Bottom corners plugged and screwed to wall

Light pelmit where used screwed to underside of wall unit

Connecting screw

Clip-on plinth

Adjustable legs may be screwed to floor

Decorative end panels where used are screwed fom inside of unit

▲ **Figure 9.34** Fixing units with adjustable legs and brackets

Purpose-made units often have provision for scribing to uneven floor and wall surfaces (see Figure 9.35). The units should be wedged up off the ground until they are plumb and level. A compass can be set to the widest gap and used to mark a line parallel with the floor. This is then cut and the operation repeated to scribe to the wall. The unit can then be screwed to battens fixed to the wall and floor. Alternatively, cover moulds can be used in place of scribing to mask any gap between the unit and an out-of-plumb wall. The cover moulds may themselves require scribing. Plastic laminate is often cut and glued to the plinth after fixing to provide a neat, easily cleaned finish.

▲ **Figure 9.35** Scribing to uneven surfaces

Partitions and screens are normally made slightly smaller than the opening in which they are to fit. They can then be packed or wedged out and suitable cover strips or linings applied to conceal the joint (see Figure 9.36).

Figure 9.37 shows how a framed dwarf screen may be fixed to the floor by inserting steel dowels into the bottom rail and grouting these into holes in the floor. Often the free end of this type of screen terminates with a post. This may be fixed to the floor in a similar way to a newel post, either by inserting a steel dowel partly into the bottom of the post and grouting it into the concrete, or, in the case of a timber floor, continuing the post through the floor and bolting it to a joist. Alternatively, L-shaped metal brackets screwed to both the framing and floor can be used to stiffen and secure the screen at an open end or door position.

Seating

Normally the manufacture of seating is more of a requirement for specialist furniture or upholstery firms than the joiner, although purpose-made joinery works may be involved with the timber frame-work for upholstered seats and other forms of continuous public seating.

required margin

overhang

compass setting equals margin plus overhang

◀ **Figure 9.36** Fixing joinery

packing

folding wedges

screen or partition

pellet

wall

recessed ply cover lining

▶ **Figure 9.37** Fixing a dwarf screen

packing

folding wedges

unit

screw fixing through folding wedges

wall

cover fillet

steel dowel grouted into floor

Banquet seating

A typical upholstered seat suitable for use in bars, hotels, office reception areas and restaurants is illustrated in Figure 9.38. This type of seating can be used on its own free standing, backing against a wall, or in pairs back to back. The basic framework to take the sprung or foam upholstery work is normally made in beech. It consists of a number of simple mortised and tenoned framed standards spaced at about 500 mm centres (the amount of space allowed for each person). The rails which tie the whole unit together are housed and screwed to these standards.

▲ **Figure 9.38** Bench seat upholstery and framework

▲ **Figure 9.39** Alternative bench seat construction

Figure 9.39 illustrates another upholstered bench seat. It can be clearly seen from the section shown that this is a more basic form of construction, where the seat and supports are framed from 45 mm × 45 mm softwood.

The seat is formed using a removable upholstered foam cushion, which is supported by 18-mm plywood. The foam backrest is upholstered

▲ **Figure 9.40** Island seating

around 12-mm plywood and is fixed through the veneered plywood back. A matching hardwood edging and capping are used to finish both the seat and back rest. These are screwed and pelleted in place. The capping may also require scribing to an irregular wall surface. The space under the seat is closed in by pinning 6-mm veneered plywood in place.

This type of seating is suitable for fixing against a wall. Alternatively, two units can be fixed back to back forming island seating as shown in Figure 9.40, in which case a wider capping would be used to cover the joint between them.

Church seating

Pews are invariably un-upholstered and little attention is paid to making them comfortable, although the use of sloping backs and seats are an improvement to straight backs and flat seats. Traditionally, they were made from oak or pitch pine, and the pew ends often decorated with ornate carvings.

A traditional style pew is illustrated in Figure 9.41. The pew ends are built up from four 45-mm boards, cross tongued together. They are housed to receive the seat back and bookshelf. The back is framed up using 45 mm × 95 mm timber and 19-mm tongued and grooved boarding, with muntins incorporated at 1 m centres to coincide with the intermediate standards. The seat, also made from 45-mm cross-tongued material, is housed into the ends, screwed and pelleted to the intermediate standards and tongued into the back panel for support. The bookshelf is also tongued into the back framing. Small brackets may be included for intermediate support at muntin positions. Both the pew ends and intermediate standards are tenoned and pinned to plinth blocks.

▶ **Figure 9.41** Traditional church pew

A similar pew, more suited to a church of modern design, is shown in Figure 9.42. Glulam construction has been used to form all of the components, which are housed and screwed and pelleted together.

45 mm glulam end

back rest/bookshelf

70 mm × 70 mm plinth block

45 mm glulam seat

housed over standards

45 mm glulam intermediate standard

▲ **Figure 9.42** Modern church pew

External seating

A traditional external **park bench** is illustrated in Figure 9.43. Typically framed up in teak, it uses pinned mortise and tenon joints. The seat slats are fixed to the end frame and intermediate cross rails with brass recessed cups and screws. The back legs have been shaped from 75 mm × 225 mm section, which gives a slope to the backrest and reduces the likelihood of the seat toppling over.

A more basic external bench is illustrated in Figure 9.44. This uses hardwood seat boards and backrest supported by pre-cast concrete standards spaced about 1.8 m apart. Intermediate standards can be introduced where longer seats are required. The hardwood boards forming the seat and backrest are coach bolted to the standards.

19 mm × 70 mm
back slats

27 mm × 70 mm
rails

45 mm × 95 mm
arm rests

pinned
mortise and
tenon

70 mm square
legs

27 mm × 70 mm
seat slats

recessed
brass cups
and screws

▲ **Figure 9.43** Park bench

45 mm × 145 mm
hardwood
boards

6 mm
coach bolts

precast concrete
standards

▲ **Figure 9.44** Bench with concrete standards

Lectern

This is a reading desk or bookstand, normally used in churches and lecture halls by readers in a standing position.

Figure 9.45 shows a modern type of lectern standing up to 1.2 m in height to the lower edge of the sloping top, which is angled at about 25°. The side support frames and cross-rails are mortised and tenoned together. The sloping top, which is fixed by screwing up through the sloping bearers, has a raised lipping on the edge nearest the reader to prevent books and papers slipping off.

Litany desk

This is similar to a lectern but lower in height, as it is intended for use either on a tabletop or on the floor of a church with the reader in a kneeling position. Figure 9.46 illustrates a litany desk consisting of two framed sides joined by a framed front panel and a sloping book top. Often the front panel is decorated with carved tracery or linen fold pattern as shown in Figure 9.47. Linen fold panels, also known as drapery panels, are intended to imitate folded or draped curtains.

front elevation

side elevation

▲ **Figure 9.45** Lectern

front elevation

▲ **Figure 9.46** Litany desk

side elevation

tracery

linenfold

▲ **Figure 9.47** Decorative panel details

Table construction

Details of a typical table suitable for use in most situations are illustrated in Figure 9.48. The table consists of four legs that are joined by four rails. The joint shown in Figure 9.49 between the rails and legs is a table haunched mortise and tenon.

The tenon is barefaced and is mitred on its end to allow for the tenon of the other rail. Alternatively, dowels, biscuits or proprietary brackets may be used for the leg-to-rail joints. Also illustrated are different methods that can be used for fixing the table top to the framework. Pocket screwing and plastic blocks are suitable for sheet material whereas the other methods allow for moisture movement when solid timber tops are used.

Laminated, veneered or solid timber top

25 mm × 150 mm rails

63 m × 63 mm tapering legs

▲ **Figure 9.48** Typical table construction

Pocket screwing or plastic block for sheet material tops

Leg

Splayed table haunch

Rail

Mitred bare-faced tenon

Button or shrinkage plate for solid timber tops

Dowels biscuits or metal bracket as alternative leg to rail joints

◀ **Figure 9.49** Table jointing details

10
Chapter Ten
Casings, false ceilings, panelling and cladding

This chapter covers the work of both the site carpenter and bench joiner. It is concerned with principles of constructing and installing casings, false ceilings, panelling and cladding. It includes the following:

- types and installation of casings
- types and installation of false ceilings
- types, manufacture and installation of panelling
- types and installation of cladding.

The term 'casing' or 'boxing' refers to the framework, cladding and trim used to form an enclosure in which service pipes are housed. The term 'cladding' describes the non-load-bearing skin or protective covering of walls and ceilings in general.

The term 'false ceiling' refers to a suspended ceiling below the structural soffit used to lower the ceiling height or conceal services.

The term 'panelling' means framing consisting of stiles, rails and muntins, mortised and tenoned together, with panels infilled between framing members. It is used to line walls and sometimes ceilings. The term is also loosely applied to wall linings made up of sheet material.

The term 'cladding' refers normally to the application of boarding, sheet materials or tiling for external protection.

Casings

Service pipework is cased or boxed in to conceal the pipes thus providing a neat, tidy appearance, which when decorated, blends with the main room decoration. In addition they must also provide access to stop valves (stop cocks), drain down valves and cleaning or rodding points or for electrical connections.

Guidelines for encasing services

In many situations encasing services is a simple process of forming an L- or U-shaped box from timber battens covered with a plywood or

Ex 32 mm × 32 mm
batten plugged and
screwed to wall
(400 mm centres)

9 mm plywood
facing nailed
or screwed
(25 mm panel pins
or 25 mm No. 8
at 150 mm centres)

Ex 25 mm × 75 mm
softwood side
screwed to batten
(400 mm centres)

Batten
screwed
to wall

Corner
batten

Face and side
screwed to
battens

Ex 32 mm × 32 mm
batten plugged and
screwed to wall
(400 mm centres)

25 mm panel
pins punched
below surface
at 150 mm
centres

25 mm No. 8
countersunk
screws at
150 mm centres

9 mm plywood
facing screwed
or nailed

Ex 25 mm × 75 mm
softwood side
screwed (38 mm No. 8)
to batten (400 mm centres)

▲ **Figure 10.1** Pipe casings

hard-board facing (see Figure 10.1). Consider the following simple rules when planning and fixing casings.

- Use standard sections of timber where possible.
- Where casing is to be tiled, ensure the dimensions are simple widths of whole or half tiles.
- Stop valves or other fittings may require access. Fit a separate length of facing board over this section.
- Use WBP (weather and boil proof) plywood or an oil-tempered hardboard for casings in wet areas, e.g. kitchens, bathrooms and laundries, etc.

Battens, typically 32 × 32 mm, may be fixed to the wall using plugs and screws or nails, cut or masonry, depending on the wall hardness.

When using 6-mm plywood or 6-mm hardboard, timber **ladder frames** (see Figure 10.2) are required for support. These are typically 25 mm × 50 mm softwood half lapped together. 9-mm and 12-mm plywood can be used for casing sides direct to battens without a supporting framework. Plywood facing can be nailed or screwed to battens/frames.

Access panels can be screwed in position using brass cups and screws typically 25 mm No. 8s. Alternatively they may be hinged as a small door.

Casings have to be scribed to wall surfaces and finished more neatly if they are to be painted or papered rather than tiled. All nails and screws should be punched or sunk below the surface ready for subsequent filling by the painter. The sharp corner arris needs to be removed with glass paper or, as an alternative, may be covered with a timber, metal or plastic trim, as shown in Figure 10.3.

Where casings are located in living rooms or bedrooms they can be packed out with fibreglass, mineral wool or polystyrene in order to quieten the noise of water passing through.

$\frac{1}{2}$ timber thickness

Fixing batten plugged and screwed or nailed to wall

Side frames screwed to battens

Frames screwed together at corner

Ladder-framed sides halving jointed and screwed

Bottom rail may be omitted if branch is near floor

▲ **Figure 10.2** Use of ladder frame for pipe casing

Sharp arris removed

18 mm panel pins

Timber corner trim

18 mm raised head screws

Metal or plastic corner trim

▲ **Figure 10.3** Alternative corner treatments

L-shaped casings

These are used for pipes in a corner:

- Mark plumb lines on walls. Use a spirit level and straightedge, as shown in Figure 10.4.
- Fix battens to the marked lines.
- Fix the side to the batten.
- Fix the front facing to the side and batten.

Alternatively;

- Make up and fix a ladder frame and fix the facings.

Where pipes branch off the main one, the side will require notching or scribing over them. For small pipes simply mark the side, drill a hole and saw the side to form a notch as shown in Figure 10.5. Larger branch pipes, are best scribed around with the face split on the pipe's centre line as shown.

U-shaped casings

These are used for pipes in the middle of a wall:

- Mark plumb lines on the wall.
- Fix battens to the marked lines.
- Fix the sides to the battens.
- Fix the front to the sides.

Air bubble central

▲ **Figure 10.4** Marking plumb line on wall

Notch for
branch pipe
drilled and
sawn out

Face panel
scribed in pieces
around large pipes

▲ **Figure 10.5** Cutting round
branch pipes

Uneven wall surfaces

Sides and faces that fit to an uneven wall surface will require scribing as
shown in Figure 10.6.

- Position the side against the wall, place a pencil on the wall surface
  and move it down to mark a parallel line on the side. Plane to the
  line, slightly undercutting to ensure a tight fit.

- Alternatively, cut the ply face oversize (say, 15 mm over required
  width).

- Position the face against the wall.

- Temporarily nail the face in position, keeping the overhang on the
  side the same distance from top to bottom.

- Set a compass or gauge slip to the width of overhang. Mark a
  parallel line on the face.

- Plane or saw to this line, slightly undercutting to ensure a tight fit.

Access panels

The edges of access panels are often chamfered as shown in Figure
10.7. This breaks the straight joint and permits a better paint finish. In
addition, removal of the panel is eased without risk of damage. Simply
remove screws, run a trimming knife along chamfered joints to cut paint
film and lift off panel. Alternatively, access panels may be hinged or
screwed into a lining and finished with an architrave trim as shown.

Horizontal pipe casings

These are usually at skirting level. They can be formed using the same
construction as vertical casings (Figure 10.8). Alternatively, they can be
formed using a skirting board fixed to a timber top as shown.

Pencil flat on wall
to scribe side

Undercut scribed
edge to ensure
a tight fit

Set compass
or gauge slip
to overhang

Line marked
with compass
or gauge slip
and pencil

▲ **Figure 10.6** Scribing and cutting to fit an uneven wall

▲ **Figure 10.7** Access panels

▲ **Figure 10.8** Horizontal pipe casings

Taller horizontal casings may have their top extended over the front facing in order to form a useful shelf:

- Mark a level line on the wall using a spirit level and straightedge.
- Mark a straight line on the floor.
- Fix battens to the marked lines.
- Fix the top and front facing.

Bath front casings

These are termed bath panels. They can be either a standard (normally plastic) set or purpose made.

Standard bath panels

Normally simply fit up under the bath rim and fix along their bottom edge to a floor batten as shown in Figure 10.9.

Read the specific instructions supplied with the panel for details.

▲ **Figure 10.9** Fixing standard bath panel

Purpose-made panels

These have to be fixed (nailed or screwed) to a batten framework as shown in Figure 10.10. This is typically made from ex 25 mm × 50 mm PAR softwood, halved and screwed together.

Halved and screwed batten framework

Wall batten

Access at tap end

Floor batten

▶ **Figure 10.10** Batten framework for a bath panel

The panels can be formed from a variety of materials. For example:

- 9-mm plywood covered with tiles.
- 9-mm plywood covered with carpet.
- 9-mm veneered plywood with applied mouldings to create a traditional panelled effect.
- Matchboarding, TG&V (tongued, grooved and vee jointed).
- Melamine-faced hardboard.

Access panels

An access panel is often formed at the tap end of the bath for maintenance purposes, rather than removing the whole panel. Figure 10.11 shows how this may be screwed in position or hinged on.

Access door

Cupped and screwed access panel

Bath rim

Set back equals thickness of panel

Panel screwed or nailed to framework

Floor batten

Halved and screwed batten framework

▲ **Figure 10.11** Bath batten framework and access panels

- Ensure the batten framework is set back far enough under the bath rim to allow for the panel thickness.
- Mark plumb lines on the walls.
- Mark a straight line on the floor.
- Fix wall and floor battens to the marked lines.
- Make and fix the battened framework.
- Fix the panel to the framework.

False ceilings

These are also known as **suspended ceilings**. They are mainly used in concrete or steel frame buildings, such as offices, shops, factories and schools, to provide a decorative finished ceiling.

The main reasons for the installation of suspended ceilings are as follows:

- to improve the thermal insulation and so reduce heat loss
- for sound insulation purposes
- for structural fire protection
- to conceal structural beams
- to route and conceal services, heating ventilation ducts and sprinkler systems, etc.
- to create integrated lighting.

In addition false ceilings may be used in houses, particularly older ones with high ceilings. This is usually to reduce the height of the ceiling, thus creating a better-proportioned room, but also to conceal an existing ceiling which may be in a bad state of repair.

Ceilings may be suspended from the underside of roof or floor structures using either a timber framework or a proprietary system.

Timber-framed suspended ceilings

A timber and plasterboard suspended ceiling is shown in Figure 10.12. This consists of ceiling joists spanning the shortest dimension of the area and skew nailed between the wall pieces which have been previously fixed at the required height. Binders are hung at intervals from the main structure to provide intermediate support to the ceiling joists and prevent them sagging. The method used to hang the binders will depend on the type of main structure and the distances between it and the suspended ceiling.

The actual ceiling finish will normally be tongued-and-grooved matchboarding, acoustic tiles or plasterboard. Where plasterboard is used, the boards should be staggered and noggins fixed between the ceiling joists to support the joints.

319

50 mm x 100 mm binder
hung from existing joists
with galvanized metal strips

50 mm x 50 mm noggins
at all board joints

50 mm x 50 mm
wall piece

50 mm x 50 mm
ceiling joists
at 450 mm
or 600 mm centres

plasterboard

hung using timber
hangers from batten
fixed to ceiling

alternative binder details

hung from steel beam
using twisted metal straps

binder

▲ **Figure 10.12** Timber and plasterboard suspended ceiling and binder details

Counter-battened suspended ceiling

This is shown in Figure 10.13. This method, which is also known as **brandering**, is used when the existing ceiling base surface is not level enough or is otherwise unsuitable for the application of the required ceiling finish. The counter-battens are fixed to the ceiling base surface, lined in and packed down as required to provide a level surface.

The ceiling finish itself can again be either tongued-and-grooved matchboarding, acoustic tiles or plasterboard. Figure 10.13 illustrates an acoustic tile finish. These tiles have a stepped tongued-and-grooved joint to enable a secret nailed fixing, although they may also be stuck in position with an impact adhesive. The joint between the wall and ceiling in this example has been masked with a plasterboard cove.

Framed panelled suspended ceilings

These may be formed in the same way as wall panelling and fixed to a supporting framework similar to that shown in the previous examples, although a more straightforward method, giving a similar appearance, is shown in Figure 10.14.

This is a *timber framed drop-in panel suspended ceiling*. It consists of polished hardwood ceiling joists spanning the shortest dimension of the

ceiling joists

25 mm x 50 mm
counter battening

25 mm x 50 mm noggins at wall

ceiling tiles

plasterboard cove

joint detail

▲ **Figure 10.13** Counter-batten suspended ceiling with acoustic tiles

binder
and
hanger

wall
piece

plywood
panel

ceiling
joist

noggin

plasterboard
panel

wall panelling

▲ **Figure 10.14** Panelled ceiling

area cut over hardwood wall pieces, which have been previously fixed
to the top of the wall panelling. Noggins are cut over and inserted
between the joists to support the panel joints. The spacing of members
is dependent on the size of panels, 650 mm centres being the most
suitable. The panels, which lay in the rebate run around the framing

▲ **Figure 10.15** Panelled ceiling assembly details

members, could be of plywood or plasterboard covered with an embossed paper or acoustic tiles. Again, binders are hung at intervals from the main structure to prevent sagging.

The advantage of this method is that it is simply pocket screwed together on site (see Figure 10.15). Also each panel may be pushed up to provide access to the services that are often run in the ceiling void. As much of the preparation work as possible should be carried out in the workshop, leaving only the wall pieces and one end of the ceiling joists to be cut on site.

Proprietary systems

There are two main metal suspended grid ceiling systems in use: the exposed grid system and the concealed Z-bar system.

Exposed grid system

In Figure 10.16, the exposed grid system is used to support either 600 mm × 1200 mm or 600 mm square plasterboard, or acoustic or other decorative infill tiles. The grid consists of a galvanised steel main T-section runners spaced at 1200 mm centres and hung on wires from the structural soffit or on straps fixed to timber joists. 1200-mm cross-tees interlock into the main runners at 600 mm centres. Where 600-mm square tiles are to be used, 600-mm cross-tees are interlocked into the centre of the 1200-mm cross-tees. An L-sectioned angle is fixed around the perimeter of the room to support and finish the edge.

The exposed parts of the grid are stove enamelled or capped in either an aluminium foil, plastic foil or a timber veneer, making a wide variety of finishes.

▲ **Figure 10.16** Exposed grid suspended ceiling

The tiles are simply dropped in place between the T-sections and secured using spring clips. A number of tiles are normally left unclipped in order to provide a convenient access to the space above. The ceiling may be given a three-dimensional effect by using rebated edge tiles.

Concealed Z-bar system

This is a concealed grid system for supporting various types of grooved-edge tiles, see Figure 10.17. The system consists of galvanised main channels which support the Z-sections. The Z-sections are clipped to the main channels at either 300 mm or 600 mm centres (depending on the tile size).

wire hangers
at 1200 mm
centres

'L' angle
fixed to wall

'Z' bars
at 300 mm
or 600 mm
centres

wire clip

main channel at
1200 mm centres

metal splice

grooved
edge tile

▲ **Figure 10.17** Concealed Z-bar suspended ceiling

The bottom flanges of the Z-section provide support for the two opposite sides of the tiles, the other sides of the tiles being joined by a metal splice or noggin tee. L-section angles are used to support the tiles and finish the ceiling where it abuts the walls.

Before fixing, the layout of the components must be planned so that the tiles are centralised and any cut tiles around the perimeter are the same width on opposite sides of the room.

Levelling

The levelling of both timber framed and proprietary suspended ceilings is extremely important, as any distortion or misalignment will be evident in its finished appearance.

To achieve a level ceiling in each case the first operation is to establish the position and fix the wall pieces. All of the other components must then be lined into these. The position of the wall pieces may be marked out by measuring up a set distance from a datum line. This will have been previously established using either a water level or an optical levelling instrument as mentioned in Book 1. However, specialist erectors working on large office or industrial contracts generally use electronic and laser levelling instruments to establish datums.

Panelling

Wall panelling is the general term given to the covering of the internal wall surface with timber or other materials to create a decorative finish. All panelling may be classified in one of two ways, either by its method of construction or by its height.

Dado-height panelling

This is panelling that extends from the floor up the walls to the windowsill level or chair-back height, i.e. about 1 m, and is known as dado panelling. See Figure 10.18.

▶ **Figure 10.18** Dado-height panelling

Three-quarter-height panelling

This type of panelling is also known as **frieze-height panelling**. It extends from the floor up the walls to the top of the door i.e. about 2 m. Traditionally a plate shelf was incorporated on top of this type of panelling to display plates and other frieze ornaments. See Figure 10.19.

Full-height panelling

This, as its name suggests, covers the whole of the wall from floor to ceiling. See figure 10.20.

Methods of construction

Traditionally all panelling was made up from solid timbers. It consisted of framing, mortised and tenoned together to receive decorative, flat or raised panels and other solid mouldings. The modern practice is to utilise sheet materials as far as possible, either in a traditional framed surround or on its own without framing to give a flush appearance.

◀ **Figure 10.19** Three-quarter-height panelling

◀ **Figure 10.20** Full-height panelling

This flush appearance can be given a traditional look by tacking planted mouldings to its surface to form mock panels. In addition to framed or sheet panelling, strip panelling made from narrow strips of boarding (T&G and V-jointed) is also used.

Vertical and horizontal sectional details of three-quarter traditional framed panelling are illustrated in Figure 10.21. Different panel and mould details have been included to show the various treatments possible. For example, panels are sometimes fitted into rebates and beaded rather than plough grooves. This method enables the polishers to work on the panels and framing before they are finally put together. Also shown are finishing details at door and window openings.

In this type of work it was common for the large architraves to be halved and mitred together at their top joint and bare faced tenoned into the back of the plinth block at the bottom (see Figure 10.22).

▲ Figure 10.21 Framed panelling details

The whole set of architraves and plinth blocks are glued and screwed together prior to fixing. The purpose of the plinth blocks is to take the knocks and abrasions at floor level that would easily damage the large moulded section architraves if they were continued down.

tops halved
and mitred

architrave

bare-faced
tenon

plinth
block

skirting
housing

rear view

scribe

internal angle

skirting

groove

tongue

▲ **Figure 10.22** Architrave, skirting and plinth block details

In addition they also ease the fixing problems and provide a neat finish when the skirting is thicker than the architrave. Skirtings were mitred in the normal way at external angles, but often scribed and tongued at internal angles and housed into the plinth block to conceal any movement.

The elevation and sectional details of a more modern design of three-quarter height panelling is shown in Figure 10.23. This is formed of sheet panelling which is fixed on the surface of a framed surround. The panels could be made of a variety of material, for example, insulation board covered with hessian or other material for good acoustics, ply or MDF with a veneer, plastic laminate, cork, lino or plastic applied finish. Other sheet materials which are suitable have either a chipboard, MDF, hardboard or ply base and a factory-finished decorative surface including melamine, simulated stonework, brickwork and vee-jointed matchboarding.

The simple framework, the bottom rail of which forms the skirting, may be painted or polished to contrast with the panel finish. Also shown is a method of forming a vertical joint for use where the total length of panelling might cause problems with handling or access.

▲ **Figure 10.23** Sheet panelling details

Whichever method of construction is used, most of the work will have been carried out off site in a joiner's shop. The panels should be delivered to site at the latest stage possible in easy-to-handle sections of up to 3.6 m in length.

Grounds

Probably the most important requirement of panelling is a straight and level surface on which to fix it. Any distortion will be exaggerated in the finished panelling and will mar its appearance. Battening out the walls with grounds can provide a straight and level surface. Grounds are normally preservative-treated softwood. They may have been framed up using halvings or mortise-and-tenon joints or alternatively supplied in lengths for use as separate grounds or counter-battening (see Figure 10.24).

The method used to fix the grounds will vary depending on the material they are being fixed to, for example, brick, block, concrete or steel, but will be either by plugging and screwing, nailing into twisted timber plugs, nailing direct to the wall with cut or hardened steel nails, or by using a cartridge fixing tool.

Uneven wall surfaces will require packings behind the grounds to achieve a flat surface. A level and straightedge can be used to test the wall in order to find any high spots. The grounds can then be fixed, packed out and lined in to this level, see Figure 10.25. As framed grounds are made up with the panelling in the joiner's shop, they will have been planned to suit each other, but when separate grounds or counter-battens are used it is essential that these are correctly positioned in order to provide the desired fixings for the panelling. These positions will be shown on the full-size setting-out rods of the panelling supplied by the joiner's shop.

Fixing panelling

The fixing of panelling to the grounds should be concealed as far as possible. There are various methods of achieving this. Figure 10.26 illustrates the following methods of concealed or 'secret' fixings:

- *Interlocking grounds.* Splayed or rebated grounds are fixed to the back of the panelling and the wall. As the panelling is lowered it is hooked in position on the grounds.

- *Slot screwing.* Keyhole-shaped slots are prepared in the back of the panelling and corresponding countersunk-head screws are driven into the grounds. The slots are then located over the projecting screws and the panelling tapped down so that the head is driven along the slot to provide a secure completely secret fixing. This method was also widely used in traditional panelling for fixing skirtings, architraves and plinth blocks.

▲ **Figure 10.24** Types of ground

▲ **Figure 10.25** Plumbing and lining grounds

▲ **Figure 10.26** Concealed fixings

- Slotted *and interlocking metal plates*. The use of keyhole slotted plates is similar to slot screwing, the plates being recessed in the back of the panelling. Interlocking plates are similar in principle to interlocking grounds, the cranked plate being fitted in corresponding positions in the back of the panelling and the face of the grounds. An alternative method is to fit the plates so that the panels can slide in sideways.
- *Pellets*. Screw-holes are counter-bored, and cross-grained pellets made from the same material as the panelling are glued and inserted in the hole. Care must be taken in matching the grain and cleaning the pellet back flush with the surface. Dome-headed pellets are also available to provide a featured fixing.

- *Cover fillets*. Panelling can be surface screwed to the grounds and then covered with a fillet pinned in place. This may be moulded to form a feature or may in fact be the skirting, cornice, frieze rail, dado rail or pilaster.

- *Nailing*. Strip panelling such as TG&V matchboarding may be secret fixed by pinning through the tongue. In other sections the nails may be partly concealed by pinning through a quirk in the moulding and filling with a matching stopping, or secret fixed through the surface by lifting a thin sliver of grain with a chisel, nailing behind this and gluing the grain back in position.

- *Metal or plastic trim*. These are mainly used to fix sheet material panelling, and in most circumstances provide a raised feature joint.

Corner details. The method of forming internal and external angles will depend on the type of panelling, but in any case they should be adequately supported by grounds fixed behind. Figure 10.27 illustrates various details. Tongue-and-groove joints, loose tongues, rebates or cover fillets and trims have been used to locate the panelling members and at the same time conceal the effects of moisture movement.

▲ **Figure 10.27** Corner details

Panel moulds. Where bolection mouldings are used, these should be slot screwed through the panel from behind. The length of the slot must be at right angles to the grain in order to allow panel movement without any possibility of splitting (see Figure 10.28). Also shown is a method used to fix planted panel moulds. They should be pinned to the framing only and not the panel. This again allows a certain amount of panel movement without splitting.

▲ **Figure 10.28** Bolection moulds and planted panel moulds

Matchboarding. Where matchboarding or similar timber strips are used for panelling, they are fixed back to the grounds using secret nails through the tongue. The boards at either end of a wall should be of equal width and may be surface nailed (see Figure 10.29).

General requirements of panelling

■ Before any panelling commences it is essential that the wall construction has dried sufficiently.

■ All timber should be of the moisture content required for the respective situation, that is, the equilibrium moisture content (MC).

■ The backs of the panelling sections should be sealed prior to fixing, thus preventing moisture absorption.

■ Timber for grounds should be preservative treated.

■ A ventilated air space is desirable between the panelling and the wall.

■ Provision must be made for a slight amount of moisture movement in both the panelling sections and the trim.

■ The positioning of the grounds must be planned to suit the panelling.

■ The fixing of the panelling to the grounds should be so designed that it is concealed as far as possible.

Secret nailed to ground

Grounds fixed to wall

Skirting finishes ends of boards to floor

End boards surface nailed and kept equal width

Secret nailed through tongue

4 3 2 1
etc

Start fixing this end first

Cover mould where used

▲ **Figure 10.29** Layout of matchboard panelling

Cladding

Cladding is the non-load bearing skin or covering of external walls for weathering purposes, for example, timber or plastic boarding, sheet material and tile hanging, see Figure 10.30. Most profiled boards are fixed horizontally, with the exception of matchboarding which is fixed vertically or diagonally. Square edged boards used for cladding are normally fixed vertically. In all situations it is essential that the cladding is positioned to shed rainwater quickly off the surface and prevent moisture penetration at joints and overlapping edges.

The Building Regulations place restrictions on the amount and location of combustible materials used to clad external walls. These restrictions are due to the risk of fire spreading from one building to another, and are more severe the closer the cladding is to the boundary. The amounts of unprotected area, which include door and window openings as well as combustible cladding, must be calculated. For example the total amount permitted for small residential buildings is around 5 m^2 at one metre from the boundary progressively increasing to no limit at six or more metres from the boundary. Cladding that has been treated with an approved exterior flame retardant may be permitted to exceed these limits, but approval from the Building Inspector will be required.

▲ **Figure 10.30** Cladding

Grounds

Cladding is normally fixed to timber battens or grounds spaced at a maximum of 600-mm centres. A moisture barrier is fixed below the cladding to battens in timber-framed buildings to provide a second line of protection to any wind-driven rain that might penetrate the cladding. This is often termed a breather paper, as it must allow the warm air vapour to pass or breath through it from inside the building and not get trapped in the wall. The moisture barrier is often omitted for claddings over brick or blockwork.

The battens are typically a minimum of 38 mm wide and at least $1\frac{1}{2}$ times the thickness of the cladding boards, are fixed to the studs of the timber frame or direct to the brick or block surface. These will run in the opposite direction to the cladding, unless counter battens are used where the first layer is fixed parallel with the cladding. They are lined and levelled to provide a flat surface in the same way as internal grounds.

Venting the cavity

Venting the space behind the cladding is important in order to reduce the risk of timber decay. Venting allows any moisture vapour passing through the wall from the inside, or any moisture absorbed by the cladding from the air to be safely dispersed, see Figure 10.31. Venting of the cavity can be considered as an increased fire hazard and cavity barriers may require fitting in order to prevent the passage of smoke and flames. Again, approval from the Building Inspector will be required.

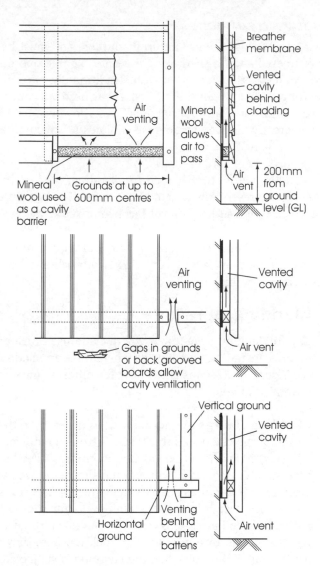

▲ **Figure 10.31** Venting the cavity

Cladding materials

- *Timber cladding* is normally specified as 16 mm thick and varies between 100 mm and 150 mm in width. Feather-edged boards will taper to about 6 mm at their thin edge. Boards are often back grooved to relieve internal stresses. This also aids ventilation of the space behind.

- *Plywood* used for cladding must be WBP rated. This means that its veneer layers have been glued using a weather- and boil-proof adhesive to give it a very high resistance to all weather conditions.

- *Fibreboard or hardboard* must be of the oil-tempered kind for weather resistance. Before fixing, hardboard sheets will require conditioning. This involves brushing up to one litre of water into the back face (rough surface) of each sheet 24 to 48 hours before fixing. The purpose of conditioning is to expand the sheet, which ensures that it dries out and shrinks on its fixings and remains flat. If this were not done it could expand after fixing resulting in a bowing or buckling of the surface.

Preservative treatment

Most softwood claddings are not naturally durable and must be preservative treated and finished using a permeable microporous paint or stain which allows the wood to breath. Traditionally, natural durable timber cladding such as Western Red Cedar was used without preservative treatment or any subsequent finish, however, as the life expectancy is greatly reduced by adverse climates it is recommended that even this is treated.

It is recommended that all timber used for cladding, the grounds as well as the boarding, is preservative treated before use. There is little point in treating the face of cladding after it has been fixed, leaving the joints or overlapping areas, back faces and grounds untreated. Any preservative-treated timber cut to size on site will require re-treatment on the freshly cut ends and edges. Applying two brush flood coats of preservative can carry this out.

Cladding details

Careful detailing is required where claddings finish around door and window openings or where different claddings join, see Figure 10.32. This is to allow for differing movement and prevent rain penetration.

Cladding fixings are normally nails at least $2\frac{1}{2}$ times in length of the cladding's thickness. Ferrous metal (metal that will rust) nails should be galvanised or sherardised to resist corrosion. Copper or aluminium nails must be used with Western Red Cedar as it accelerates rusting in ferrous metals and causes unsightly timber staining. Boards should be single nailed to each batten and care must be taken to ensure the nails do not go through the board below where they overlap, as splitting will occur when the cladding is subjected to moisture movement, see Figure 10.33.

Corners   **Frames**

Vertical grounds

Mastic seal

Window trim

Corner trim

Horizontal cladding

DPC material tacked to back of window frame

Plywood cladding

Joint filled with sealant

Cladding

Ground

Vent

Flashing

Solid 'L' corner trim

Groove in back of trim to prevent capilarity

Joint cover trim

Sills

Flashing tacked to underside of window sill

Vent

Vented cavity

Cladding

Ground

▲ **Figure 10.32** Typical cladding details

▶ **Figure 10.33** Typical cladding fixings

Nails towards bottom edge but avoiding overlap

Packer to tilt bottom board

Nails towards bottom of each board

Nails through centres of board

Feather edge   **Shiplap**   **Board on board**

Temporary construction and site works

This chapter covers the work of the site carpenter, which is mainly of a temporary nature and includes the following:

- centring to support arched brickwork under construction
- shoring to support brickwork that is either failing or being extensively altered
- formwork to shape and support wet concrete until it cures sufficiently to become self-supporting
- formwork for walls, columns, slabs and beams
- timbering to support the sides of trench excavations whilst foundational work is being carried out
- hoardings and fencing posts
- inspection and safety aspects.

Centring

To enable bricklayers to build an arch they require a temporary framework, which is normally made of timber and is known as centring.

There are two main types of supporting framework.

Turning piece

This is the simplest form of centre and is cut from a solid piece of timber. It is only suitable for flat or segmental arches, which are of low rise, approximately 150 mm from the springing line (see Figure 11.1).

Built-up centres

These are used where the rise exceeds 150 mm. Centres for low rises up to approximately 200 mm can have solid ribs, as illustrated in Figure 11.2. For rises in excess of this, made-up or laminated rib frames are used. Figure 11.3 shows a built-up centre with rib frames made up from a double layer of separate ribs held together at the bottom with a tie. This is suitable for rises of approximately 500 mm.

▲ **Figure 11.1** Turning piece

▲ **Figure 11.2** Centre for low rise

▲ **Figure 11.3** Centre with rib frames

Where the width of the brickwork exceeds half a brick, cross-braces should be used to stiffen the centre. The two sets of ribs are spanned around the centre by lagging and at the bottom by bearers. The laggings and bearers should be 20 mm shorter than the depth of the arch, thus ensuring that the centre will not obstruct the bricklayer in levelling and plumbing the brickwork. Timber battens or plywood may be used for the lagging, depending on the type of work in hand. In general plywood or closed timber lagging is used for brickwork and open timber laggings for stonework.

Gothic

Figure 11.4 shows a centre for an equilateral or Gothic arch. Each rib frame is made from two separate ribs, held together at top and bottom with a tie and a strut in the centre for additional support. Cross-braces are fixed to the struts to stiffen the centre.

Solid ribs

Figure 11.5 shows a centre made up using two solid ribs cut from a sheet of 18-mm plywood. Laggings join the ribs around the curve

▲ **Figure 11.4** Centre for Gothic arch

▶ **Figure 11.5** Centre with two solid plywood ribs

338

and bearers at the bottom. The vertical and the horizontal timber stiffeners have been included to provide rigidity. These take the place of the cross-braces used in traditional centres.

Large spans

The previous turning pieces and centres were suitable for arches with spans of up to approximately 1 m. Centres for larger spans can be made using the same form of construction, the only difference being that the size of the timber used should be increased to support the additional load.

Figure 11.6 shows a trussed arch centre, which is suitable for spans of up to 3 m. The method of making up this centre is very similar to a trussed rafter. The components are butt-jointed together and secured by nailing plywood gussets or nail plates on either side. Before assembly the timber used should be prepared and individual ribs pre-cut to a template.

Arches with spans in excess of approximately 3 m are rarely encountered in normal building work. They are normally restricted to civil engineering work such as bridge or tunnel construction.

▲ **Figure 11.6** Modern arch centre

Semi-elliptical

Figure 11.7 shows a semi-elliptical arch centre made up using the trussed gusset plate method. This method can be used for making up any arch centre irrespective of its outline shape.

▲ **Figure 11.7** Semi-elliptical modern arch centre

Setting out and assembly

Segmental centres and turning pieces are set out using either the triangular frame method or a radius rod. Where the radius rod is used, the required radius will have to be calculated or the centre point found by bisection. Other arched centres are normally set out using a radius rod or trammel. The geometry required for setting out these and various other arch centre outlines is covered in Book 1.

When making the outline of an arch an allowance must be made for the thickness of the laggings.

The rib frames for built-up centres should be assembled over a full-size outline of the centre that has been marked either on a sheet of ply or the workshop floor. This ensures the ribs are made up to the required shape. Figure 11.8 shows one rib frame for a small semicircular centre. The inner ribs, outer ribs, struts and tie are clench nailed together. The outline of the arch has been marked and the rib frame is ready to be cut on the band saw. Once the rib frames have been assembled the lagging bearers and braces can be nailed in position.

On site, where a band saw may not be available, a portable power jigsaw can be used. The ribs for larger centres are normally pre-cut to a template before the rib frames are assembled.

outline of arch marked out
on ribs ready for cutting

pencil

bradawl

radius rod

▲ **Figure 11.8** Marking out semicircular arch

Support of centres

Turning pieces and centres can be supported, levelled and adjusted by means of either adjustable steel props or timber posts and folding wedges. The props or posts transfer the load imposed on the centre to the ground or other support: the adjustment given by the props or wedges enables the centre to be levelled and in addition facilitates the easing and striking (slight lowering and removal) of the centre without damage to the finished brickwork. Large-span centres require additional intermediate props in order to support their imposed load. Once the arch has been built and the mortar joints have set sufficiently, the centre can be removed; but before this, props or wedges should be gradually eased until the arch takes up the load of the structure.

Shoring

When a building is to be extensively altered or is considered to be structurally unsound, possibly owing to ground subsidence, it may be necessary to give temporary support to the walls of the building. This temporary support is called shoring. The three main types used are illustrated in Figure 11.9, these being:

- vertical shores
- raking shores
- horizontal shores.

Vertical shores give vertical support, horizontal shores provide horizontal support, whereas raking shores fall somewhere between the two, providing a certain amount of both vertical and horizontal support. Often combinations of two or more types are required.

As an alternative to timber, tubular scaffolding or proprietary steel systems are often used. The overview in the following section is restricted to timber.

vertical   raking

horizontal

▲ **Figure 11.9** Types of shoring

Vertical shores

The purpose of vertical shores is to support the superimposed or dead loads of a section of a building during alteration or repair work. A typical situation where vertical shores are used would be where an opening is to be made in a load bearing wall. The wall above the opening must be supported while a lintel is inserted and the wall made good around the lintel. Figure 11.10 shows a typical detail for vertical shores.

▶ **Figure 11.10** Detail for vertical shores

Sequence of operations

1. Strut all window openings. This is done by placing lengths of 50 mm × 75 mm timber in the window reveals and driving struts tightly between them.
2. Prop floors using 75 mm × 225 mm head and sole plates and adjustable steel props. On hollow ground floors, some boards should be removed so that the prop can extend down to the oversite concrete.
3. Cut holes in the wall for the needles. These should be spaced at 1.2 m centres under solid brickwork and not placed under windows, etc.

4. Place the sole plates on firm ground both inside and outside the building, as these are going to spread the load. It is a good idea to place the lintel in position on the floor before the shoring is erected, as access may be restricted afterwards.

5. Insert the needles through the wall and support them on dead shores. Drive hardwood-folding wedges between the vertical shores and the sole plates to bring the needles tight up under the brickwork. Packing may be needed between the needles and the brickwork. Where cement mortar is used for the packing it must be allowed to harden before any load is placed upon it.

6. Securely fix all joints between the sole plates, the vertical shores and the needles with steel dogs.

7. Cross-brace the vertical shores to prevent any possibility of them overturning.

8. Check all of the shoring for tightness, i.e. props, wedges, struts, bracing, etc., and adjust if required.

9. Remove wall and build piers and padstones to support the lintel.

10. Insert the lintel. This is usually either a universal steel beam or a concrete lintel.

11. Make good the brickwork around the lintel and allow at least seven days for this to thoroughly set.

12. Ease and remove the shoring and make good the holes in the wall through which the needles passed.

The easing entails partially slackening the wedges. After about one hour these should be checked. If they are still loose then all of the shoring can be removed using the reverse order to that in which it was erected. But if the wedges have tightened, some form of settlement has occurred. The new brickwork around the lintel should be checked for any signs of cracking. If this is satisfactory then only some initial settlement has taken place and the easing procedure can be repeated.

Timber sizes

The size of timbers used for vertical shores will depend on the building to be supported, but an approximate method of determining a suitable section of timber for a two-storey building is to divide the shore's height by twenty. For example, vertical shores which are 3 m high require a section of 3000 ÷ 20 = 150 mm. Therefore the section required is 150 mm × 150 mm.

Where the building exceeds two storeys, 50 mm should be added to the section for each additional storey. For example, vertical shores which are 3 m high for a three-storey building require a section of 150 mm + 50 mm. Therefore the section required is 200 mm × 200 mm.

In order to save on costs, second-hand timber can be used providing it is sound and free from defects. Where large-section timber is not available, this can be built up or laminated from smaller sections; for example, three lengths of 50 mm × 150 mm bolted together will give a 150 mm × 150 mm section.

Raking shores

These are shores that are inclined to the face of a building. Their purpose is to provide support to a building that is in danger of collapsing, or they may be used in conjunction with vertical shores for support during alteration or repair work.

In order to provide maximum support to the floors and to prevent 'pushing in' defective brickwork, the tops of the rakers must be positioned in relation to the floors (see Figure 11.11). When the joists are at right angles to the wall the centre line of the raker should, when extended, intersect with the centre of the joist's bearing. For joists that are parallel to the wall, the centre line of the floor, wall and raker should intersect.

Single raking shore

Figure 11.12 shows the front and side elevation of a single raking shore. These should be positioned at each end of the building and over intermediate cross-walls at between 2.5 m and 5 m apart, depending on the building's condition.

The term 'single raking shore' does not refer to the number of shores used but to the number of rakers in each shore, single, double, triple, etc.

Figure 11.13 shows the jointing arrangement at the top of the raker. The raker is cut to fit the wall piece and notched around the needle. This locates it and prevents lateral movement. A straight-grained hardwood must be used for the needle, which is mortised through the wall piece and allowed to project into the brickwork by about 100 mm. The function of the wall piece is to spread the thrust of the raker evenly over a large area of the wall. In order to do this effectively it must be securely fixed back to the wall using wall hooks. The cleat, which is housed into the wall piece, stiffens the needle.

▲ **Figure 11.11** Position of raking shores

▲ **Figure 11.12** Raking shore elevation

▲ **Figure 11.13** Joinery arrangement at top of raker

▶ **Figure 11.14** Raker detail at ground level

Figure 11.14 shows the detail at ground level. This consists of the raker bearing on the sole plate at an angle of about 80–85°. This angle must always be less than 90° to enable the raker to be tightened when it is levered forward.

Where the ground is weak, a grid or grillage of sleepers can be placed under the sole plate to spread the load over a larger area.

Double raking shores

Figure 11.15 shows a double raking shore to a two-storey building. The head of each raker must be positioned in relation to the floor as before. In any shore system the internal angles between the outer raker and the horizontal should be between 60° and 75°.

▶ **Figure 11.15** Double raking shore to two-storey building

Multiple raking shores

Figure 11.16 shows a multiple raking shore to a four-storey building. Where a sufficient length of timber is not available for the outer raker it may be made up using two pieces. The longest available length is used

wall piece
wall hook
cleat
needle

braces

rider shore

folding wedges

back shore

dog   cleat

sole plate
and grillage

◀ **Figure 11.16** Multiple raking
shore

for the top rider shore. The remaining length is made up with a
backshore that lays against the adjacent raker. Adjustment is made by
tightening the folding wedges between the rider and backshore.

Sequence of operations

1.   Strut windows as before.
2.   Cut out half-bricks in the wall to accommodate the needles.
3.   Fix the needle and the cleat into the wall piece.
4.   Place the prepared wall piece up against the wall and fix.
5.   Prepare the sole plate and grillage.
6.   Place the raker in position and tighten using a nail bar to lever it
     forward on the sole plate.

Note that over-tightening can be dangerous. The shore is there for
support only. No attempt should be made to push back bulging
brickwork.

7.   Fix the foot of the raker with a cleat and dogs.
8.   Brace the raker back to the wall piece. Diagonal cross-braces and
     ties should be used between adjacent shores.
9.   Check the shoring for tightness periodically and adjust as
     required.

Timber sizes again depend on the buildings being supported, but as a
guide 150 mm × 150 mm rakers are suitable for single-shore systems,

175 mm × 175 mm for double and 225 mm × 225 mm for triple. As before, these sections can be laminated by bolting a number of smaller sections together.

Horizontal shores

These are also known as flying shores and are used to provide horizontal support between the walls of buildings that run parallel to each other. They are mainly used for support when an intermediate building of a terrace has to be demolished prior to rebuilding, or across a narrow street or alley way, to support a weak wall off a sound one. In these situations the use of raking shores may be impracticable because of the limited space available or because of the obstruction they would cause at ground level.

Buildings of two storeys in height require only a single horizontal shore suitably strutted as shown in Figure 11.17. For taller buildings multiple horizontal shores may be required. Figure 11.18 shows the arrangement of a double horizontal shore.

▲ **Figure 11.18** Double horizontal shore

▲ **Figure 11.17** Single horizontal shore

▲ **Figure 11.19** Centre-line intersection for horizontal shores

The arrangement and position of the shores, wall pieces, needles and cleats are similar to those used for raking shores.

Figure 11.19 shows the centre-line intersections of floor, wall and shoring timbers. The positions of needles and cleats are inverted to support the horizontal shore and take the downward thrust from the lower rakers. Hardwood folding wedges are used to tighten the shore. Enlarged details are shown in Figure 11.20.

▲ **Figure 11.20** Enlarged details of intersections

Often more than one horizontal shore is required to provide sufficient support, in which case a system of shores can be used with individual sets of shores positioned near the ends of the opposite walls and at 2.5 to 5 m intervals.

Figure 11.21 illustrates the use of a horizontal shore where one building is higher than the other. The floor levels of the two opposing buildings do not always coincide. In these cases the floor levels of the weaker building should be used to determine the positions of the shores. A thicker wall piece is used on the sound building to transfer the thrust to the floors and reduce the risk of 'pushing in' the wall.

Sequence of operations

1. Strut window opening (if any).
2. Cut out half-bricks in the wall to accommodate the needles.
3. Fix the needles and cleats into the wall pieces.
4. Place the prepared wall pieces up against the walls and fix.
5. Lay horizontal shores and posts in position and wedge up.
6. Place rakers and straining pieces in position and wedge up.
7. Nail all wedges and dog joints on completion to prevent lateral movement.
8. Fix diagonal cross-braces between adjacent sets of shores.
9. Check the shoring for tightness periodically and adjust by re-driving wedges if required.

▲ **Figure 11.21** Horizontal shore for different floor levels

Table 11.1 Sizes for horizontal shore members

Span	Horizontal shore	Raker posts	Straining piece	Wall piece
up to 4 m	100 mm × 150 mm	100 mm × 100 mm	50 mm × 100 mm	50 mm × 225 mm
4 m to 6 m	150 mm × 150 mm	100 mm × 100 mm	50 mm × 100 mm	50 mm × 225 mm
6 m to 9 m	150 mm × 225 mm	100 mm × 150 mm	50 mm × 100 mm	75 mm × 275 mm
9 m to 12 m	225 mm × 225 mm	150 mm × 150 mm	50 mm × 150 mm	75 mm × 275 mm

Timber sizes in this case depend on the span of the horizontal shore and not on the height of the building. Typical sizes for members are given in Table 11.1. These can also be laminated by bolting a number of smaller sections together as before.

Shoring safety

Specialist firms normally carry out shoring. However, in all cases the work must be designed and closely supervised during erection by an experienced person with extensive knowledge of shoring operations.

Inspections

Prior to erection a detailed inspection of the building should be made to determine its structural condition. Contact the local authority and service undertakings to determine the locations of underground sewers, water, gas and electricity mains and telephone cables. These must be avoided when positioning the sole plates of vertical and raking shores.

Irrespective of the type of shoring to be used all materials and component parts must be thoroughly inspected before erection. Anything that is found to be defective must be clearly marked that it is unsafe for use. Timber should be discarded. Metal parts can often be returned to the manufacturer for straightening or repair.

After erection, every shoring system must be inspected at seven-day intervals and after adverse weather conditions for signs of distortion and movement.

Points to check are:

■ Distortion of members caused by overloading.
■ Tightness of members. Slackness can be caused by timber drying out, or by soil movement due to repetitive freezing and thawing or soaking and drying of the ground.
■ Displacement of timbers including wedges, possibly due to traffic vibration, moving plant or settlement of building.
■ If further movement, cracking or bulging of the building is noticed, the designer of the shoring should be contacted immediately for advice, as such movement could possibly end in total collapse.

Removal of shoring

Horizontal or raking shores supporting a structurally unsound building cannot be removed until the defect has been remedied (walls rebuilt or foundations underpinned, etc.) or the building has been completely demolished.

The procedure for striking is normally the reverse of that used for erection. This should be done in easy stages over several days so that the structure gradually takes up its full load. Look out for signs of building movement such as cracks in the brickwork or the opening up of mortar joints. If such movement is found, consult a structural specialist immediately.

Raking or horizontal shores should be the first to be erected and the last to be removed when used in conjunction with vertical shores.

Formwork

Formwork can be defined as a temporary structure which is designed to shape and support wet concrete until it cures sufficiently to become self-supporting, and thus includes not only the actual materials in contact with the concrete but all of the associated supporting structure.

Glossary of formwork terms and basic principles

Refer to Figures 11.22 to 11.46

- *Adjustable prop.* A proprietary steel prop with provision for adjustment so that its length may be varied (Figure 11.22).
- *Beam box.* The formwork for a beam, which includes the sides and the soffit.
- *Beam cramp.* A proprietary cramp or timber yoke which maintains a constant beam width during casting (Figure 11.23).
- *Blow hole.* A small hole or cavity in the concrete face caused by an air pocket, normally not more than 15 mm in diameter.
- *Box-out.* The formwork for a pocket or opening in the concrete item, e.g. bolt hole, service duct opening, door opening, etc. (Figure 11.24).
- *Column box.* The formwork for a column.

fork head fitting

nail fixings   ledger

wedge

fork head must be concentrically loaded

head plate flat and welds intact

high-tensile steel pin on chain (no substitutes permitted)

prop to be plumb, and not visibly bent, dented or pitted

base plate flat, welds intact and centred on sole plate

adjustable   fixed

▲ **Figure 11.23** Beam cramp

▲ **Figure 11.22** Adjustable prop

alternate corners lapped
(chasing corners)

corner
block

box can be
folded out
after removing
corner blocks

ply or
timber
sides

central stiffener
required for large
box outs

box out for
service duct
opening

chasing corners

6 mm ply sides

sides
pinned

tapering box-out
for bolt pocket

▶ **Figure 11.24** Box-outs

- *Column cramp.* A proprietary steel cramp or timber yoke which holds the column box tightly together during casting (Figure 11.25).
- *Cover.* The measurement between the concrete surface and the reinforcement. This protects the reinforcement from corrosion and increases the resistance to fire (Figure 11.26).
- *Cube test.* A number of standard-size concrete cubes are cast on site from specified mixes. After curing one is placed in a press and subjected to an increasing load to determine its crushing or compressive strength. After curing for twenty-four hours in the metal moulds, the cubes are stored in water until they are sent to the laboratory for testing, normally at seven and twenty-eight days after casting.
- *Decking.* The sheeting or soffit of slab (floor and roof) formwork.
- *Distance piece.* A short member used to ensure the constant spacing of parallel wall and beam formwork.
- *Double-headed nail.* A wire nail with two heads used for formwork. The first head is driven home for strength, the second is left projecting, enabling easy withdrawal.
- *Draw.* The lead or taper of a fillet to allow its ease of removal from the cast concrete. Large features may require a false lead or draw (Figure 11.27).
- *Edge form.* the vertical formwork to the edge of a slab, path or road, etc. (Figure 11.28).

▲ **Figure 11.25** Column cramps

200 mm	
500 mm	
450 mm	2.800m high column
400 mm	
350 mm	
300 mm	
250 mm	
200 mm	
150 mm	

typical spacings

Stirrups or links

Main bar

Cover

Cover

Insufficient cover results in steel expansion and spalling concrete

◄ **Figure 11.26** Concrete cover to reinforcement

◄ **Figure 11.27** Draw

tapered sides ease striking

square feature false draw

this screw removed before striking

▲ **Figure 11.28** Road edge form

- *Expanded metal.* An open-mesh sheet metal which is useful for forming box-outs and permanent stop ends. It is fairly easy to cut using shears. Reinforcing bars can be pushed through without difficulty. It has the advantage, when used as a stop end, that it provides a key for the subsequent pour.

- *Expanded polystyrene.* A rigid plastic, obtainable in blocks or sheets. Its main use is for forming box-outs and other small pockets for holding down bolts, metal handrails, and balusters. After use, the polystyrene can be removed by breaking it up, burning it out, or softening it out with petrol.

- *Fair face.* A high-quality plain concrete finish achieved straight from the form, without any subsequent making good or touching up.

- *False work.* The temporary support structure that supports the forms. Normally only used to describe the support structure for bridges and other major civil engineering works.

- *Fillet.* a small section of material fixed to the form face in order to produce a chamfer, bullnose or groove, etc. (Figure 11.29).

▶ **Figure 11.29** Fillet

▲ **Figure 11.30** Floor centre

- *Forkhead.* A fitting located at the top of adjustable steel props when used for decking, to provide a secure bearing for ledgers.

- *Floor centre.* A proprietary metal beam, normally of a lattice or box design, which is adjustable in length and used in place of ledgers in soffit formwork (Figure 11.30).

- *Form filler.* Filler used to repair surface damage to sheathing materials and fill small gaps, nail and screw holes for fair-faced work. Repairs are normally overfilled and sanded down flush once the filler has hardened.

- *Form lining.* The actual material used to line the inside of a form in order to produce the desired finish.

◄ **Figure 11.31** Form panels

▲ **Figure 11.32** Use of form tape

■ *Form panel*. A standard panel either of a proprietary metal, or of a framed design, or specifically made up using a timber frame and ply sheathing, normally based on standard sheet size. Prefabricated form panels can be used for a wide range of work including wall, slab construction and as baseboards for pre-cast work (Figure 11.31).

■ *Form tape*. A waterproof adhesive tape used to seal joints in the form face, thus preventing grout leakage (Figure 11.32).

■ *Form ties*. A bolt or tie rod, often called a wall tie, used to hold the retaining sides of formwork in the correct location. The main types in current use are coil tie, snap tie, she-bolt and through tie (Figure 11.33).

■ *Gang mould*. A pre-cast mould to cast more than one unit in the same casting (Figure 11.34).

■ *Grout loss*. The leakage at form joints of a mixture of cement, sand and water.

■ *Hydrostatic pressure*. The pressure exerted on the formwork by the liquid concrete.

■ *Insert*. A dovetail timber block or other proprietary item which is cast in the concrete to provide a subsequent fixing (Figure 11.35).

■ *In situ cast concrete*. Where the product is cast in the actual location or position required.

■ *Joist*. Normally 75 mm × 100 mm timbers used to support the decking material for concrete floor slabs. Proprietary aluminium joists are also available (Figure 11.36).

loop coil tie
for single-face
or climbing
formwork

baffle prevents
water seepage

water bar coil tie
for water-retaining
structures

coil tie

plate washer

coil tie bolt

cone

coil tie
(remains in
wall)

snap tie

break-back point
(after striking)

wedge

cone

she-bolt

welded water
bar tie bolt
for water
retaining
structures

waler plate

nut

she-bolt withdrawn
by turning square
shank of bolt

she-bolt

tie bolt
(remains
in wall)

through tie

cone and tube
may be used
for wall spacing
and to ease tie
rod withdrawal

plate washer

nut

threaded tie
rod

greased to
ease withdrawal

▶ **Figure 11.33** Form ties

screw to hold
centre division

25 mm x 50 mm
distance piece

18 mm plywood
ends and cleats

centre division

wedges

18 mm plywood
sides and base

50 mm x 75 mm
bearers

25 mm x 50 mm
batten

nut, bolt and washer

▶ **Figure 11.34** Gang mould

Bend nail over
after striking for
safety

Nail or screw
fixing to formface

Cast-in
threaded
socket

Bolt through
formface

Bolt through formface
Remove before striking

Cast-in
fixing lug

Column
corner
guard

Polystyrene fill

Seal to prevent
grout infiltration

Dovetail socket, provides
fixing for brick or wall ties

◀ **Figure 11.35** Formwork inserts

decking

timber insert
to fix decking

joist
secondary
beam

ledger
primary
beam

adjustable steel
prop or other
support system

◀ **Figure 11.36** Proprietary
aluminium joints and ledgers

- *Key*. A small indent or exposure of concrete surface to provide a key
  for the next casting.

- *Kicker*. A small upstand of concrete, the same plan size as the
  finished wall or column but only up to 150 mm in height. Its

▶ **Figure 11.37** Kicker

▲ **Figure 11.38** Lacing

purpose is to locate the bottom of the form and prevent grout loss (Figure 11.37).

- *Lacing*. Horizontal members used to space and tie together vertical supports. Their use converts adjustable steel props into a support system, thus greatly increasing their safe working load (Figure 11.38).

- *Ledger*. Normally 75 mm × 150 mm timbers, supported by adjustable steel props which in turn support the joists for concrete floor slabs. Also known as a runner. Proprietary aluminium ledgers are also available (Figure 11.39).

▲ **Figure 11.39** Timber ledger and joists

356

- *Lift.* The height of concrete that is poured in one operation.
- *Live load.* The temporary load imposed on the formwork by workmen and concreting equipment.
- *Mould.* The formwork for a pre-cast item.
- *Permanent form.* Formwork or the form face left permanently in position after casting.
- *Pre-cast concrete.* This is where the product is cast out of location, on site or in a factory.
- *Release agent.* A substance applied to the formwork to prevent adhesion between the form face and the concrete surface.
- *Ribbon.* A horizontal member fixed to a soffit, its purpose being to prevent beam sides and slab edges spreading. May be used to strut off (Figure 11.40).
- *Sheathing.* The face of vertical forms.
- *Shuttering.* A term sometimes used when referring to formwork for in situ concrete structures.
- *Sill.* A horizontal member, also known as a sole plate, used under vertical supports to evenly spread the load.
- *Slump test.* An on-site test to measure the workability of concrete and compare its consistency between successive mixes of concrete. A metal slump cone is filled with concrete. After compaction the cone is removed and the slump measured. The workability of the mix increases with the slump.
- *Soffit.* The underside of a floor slab or beam; soffit formwork is known as decking (Figure 11.41).
- *Soldier.* A vertical member to stiffen forms. Normally used in pairs in conjunction with form ties (Figure 11.42).
- *Stop end.* A sheathing member used to form a construction or day joint at the end of a day's concreting.
- *Strike.* To strike or striking is the removal of formwork after it has cured sufficiently to become self-supporting.
- *Strong back.* a proprietary soldier normally of a substantial metal construction and used where high pressures are anticipated (Figure 11.43).

▲ **Figure 11.40** Ribbon

▲ **Figure 11.41** Soffit

▲ **Figure 11.42** Soldier

▲ **Figure 11.43** Strong back

special crane attachment
for lifting tables

to crane

adjustment for
levelling and striking

castors fit into base for moving tables

▲ **Figure 11.44** Table form

▲ **Figure 11.45** Waffle form

▲ **Figure 11.46** Waling

■ *Table form.* A proprietary decking and support system for soffits in the form of a table which can be struck, crane handled and repositioned in one piece (Figure 11.44).

■ *Trough form.* A hollow form used to produce trough-shaped recesses in the concrete soffit. Often made of glass-reinforced plastic.

■ *Vibrator.* A mechanical tool that imparts vibrations into the concrete to assist in its compaction. Types include an immersion vibrator, known as a poker, or an external vibrator attached to the outside of the formwork.

■ *Waffle form.* A domed hollow form, square on plan, used to make recesses in the concrete soffit. Normally made of glass-reinforced plastic (Figure 11.45).

■ *Waling.* A horizontal member to stiffen forms, and used in conjunction with soldiers or strong backs and wall ties (Figure 11.46).

■ *Yoke.* The arrangement of members to encircle and secure a form, thus preventing movement. For column formwork these have now been superseded by proprietary metal column cramps.

Materials for formwork

When choosing the materials for formwork the following points should be born in mind:

■ strength of materials;

■ economic use of materials;

■ ease of handling, making and erecting;

■ facilities for adjustment, levelling, easing and striking;

■ the quality of the finish required.

Timber

Softwood is the most commonly used material. The main reasons for this are:

- availability
- economy
- structural properties
- ease in working and handling.

V-grade European redwood, European whitewood and Western hemlock are the species most widely used. For heavily loaded members the extra cost of using strength-graded timber can often be justified.

It is common practice to use PAR timber, not only because of its uniform cross-section in any one parcel but also because it is better to handle and easier to clean off after any concrete spillage or seepage.

The use of solid timber in formwork, for both quality of finish and economic reasons, is almost exclusively restricted to the support and framing members and rarely the actual soffit or sheathing. The exception is where decorative board-marked feature finishes are required, in which case the formwork is lined with either prepared or sawn softwood boards.

A prior soaking in water or a sandblasting treatment may enhance the natural grain pattern that is transferred to the finished concrete.

The reuse potential of timber-boarded formwork is between one and ten times, depending on the quality of finish required and the care taken in striking and handling.

Plywood

The two main types of plywood in common use are Douglas fir and birch. The grain pattern of Douglas fir plywood will be transferred to the concrete and can be used as a decorative feature, whereas birch plywood leaves a much smoother finish to the concrete.

Plywood for concrete formwork must be exterior quality. A/C or GIS grades are normally specified if a smooth finish to the concrete is required. B/C, BB/C or select sheathing is used where minor surface irregularities in the concrete finish are permissible. C/C or sheathing may be specified for underground or covered work where the concrete finish is not important.

Surfaced plywoods are used almost exclusively for high-quality repetitive work. This gives a greater number of reuses and a smoother concrete finish, and also reduces the effects of moisture and abrasion. These factory-applied surface treatments include:

- barrier paints and varnishes;
- cellulose film surface, impregnated with a phenolic resin;
- overlay of glass fibre impregnated with polyester or phenolic resin (GRP overlay).

On site, any cut edges or holes must be sealed with a barrier paint or other waterproofing agent to prevent moisture absorption into the exposed layers, which would result in swelling and subsequent delamination.

Care in the handling of all types of plywood is essential, but more especially with surfaced plywoods as damaged surfaces are not so easily repaired.

The reuse potential of plywood varies widely depending on its handling and surface treatment. Typical reuse values for plywood, assuming reasonable care in handling, are given in Table 11.2.

Table 11.2 Reuse values for plywood formwork

Type	Reuses (up to)
Untreated	10
Treated with barrier paint or varnish	20
Resin film surface	50
GRP overlay	100

Profiled or textured surface plywoods are also obtainable. These impart a featured surface to the concrete. The reuse potential of these textured panels is limited as they become progressively more difficult to clean and maintain after each use.

As plywood is a laminated construction it has more layers parallel to the face grain than across it. This makes it stronger parallel to the face grain. Therefore for maximum strength the face grain should be placed at right angles to the panel supports. Where this is impracticable extra supports can be used.

Chipboard

Only exterior-grade chipboard or special-quality boards should be used as a structural sheathing material. Standard grades are unsuitable as they are not sufficiently resistant to moisture absorption; this leads to a large amount of moisture movement and a substantial loss in strength.

Special film-faced chipboards for formwork are also available. These have a fairly high reuse potential if properly treated and handled. In general, the number of reuses that can be expected from chipboard is less than that from the equivalent grade of plywood.

Fibreboard

This is used as a form lining material, and is normally nailed or glued to a supporting backing of timber or plywood. The two main qualities used for formwork are standard hardboard and oil-tempered hardboard.

The oil-tempered boards have a smoother surface and are less absorbent. They produce a better concrete finish. The oil-tempered boards are also stronger and stiffer but tend to be brittle; therefore the standard boards are more suitable for curved formwork as they can be bent to a smaller radius.

Hardboard's reuse potential is fairly limited. Standard boards are considered suitable for one use only, whereas oil-tempered boards if properly treated may give up to 10 uses. However, the quality of the concrete finish will rapidly deteriorate with each reuse.

Other form lining materials

A variety of other materials is available for lining forms. These include building paper, corrugated cardboard, plastic laminate, moulded rubber, expanded polystyrene, etc. They are applied to the face of the form in order to obtain a specific concrete surface finish, for example, smooth, textured or patterned.

Medium-density and softboards. These are sometimes used as absorbent form liners, thus reducing the risk of blow holes in the concrete finish and at the same time creating a fine surface texture. These boards are only suitable for one use.

Permanent formwork

Woodwool slabs are used as permanent formwork sheathing or decking for walls and flat floor slabs in addition to permanent waffle and trough forms. The open texture of the slabs provides an ideal key for the concrete and also any subsequent finish applied to the wall or ceiling. The use of woodwool slabs has the added advantage that they have good thermal, acoustic and fire-resisting characteristics.

Metal formwork

Steel and aluminium are normally used for formwork in circumstances where there is a high degree of repetitive work. The two main types of metal formwork available are:

- proprietary forms
- special purpose-made forms.

Proprietary forms

These consist of a wide range of equipment, including the following:

- Steel-framed panels with plywood or metal sheathing which may be used for walls, columns and slabs.
- Aluminium beams for use as ledgers, joists and walings.
- Adjustable props, table forms and other support systems.
- Column cramps, beam cramps, wall ties and a variety of other ironmongery designed to secure formwork.

These proprietary systems are available for sale or hire from a number of specialist formwork suppliers.

The main advantages in using metal forms are their high strength, accuracy, smooth concrete finish, ease of erection and economy for repetitive use. Sometimes they have the disadvantages of being not readily adaptable, not readily acceptable of inserts, box-outs and fixing blocks, and generally heavier than other forms, thus requiring crane handling.

Special purpose-made forms

These are normally manufactured for one specific job. They tend to be used more in civil engineering rather than construction work. Typical applications include tunnelling, retaining walls, sea walls and slip forming, etc. In addition the use of special purpose-made forms may be considered where a large number of identical components is required, for example, pre-cast moulds, column forms or complicated shapes such as mushroom-headed columns, etc.

The reuse potential of metal forms is very high, and provided they are handled with care and thoroughly cleaned, oiled and maintained after each use, between 100 and 600 reuses can be expected. Where the forms have become damaged, distorted or otherwise defective, for both accuracy and safety reasons they should not be used. As the repair of these forms is a highly specialised operation they should either be discarded or returned to the manufacturer for repair or replacement.

Plastic formwork

Sheet or foamed plastic can be an economical formwork material where repetitive use, high-quality finish and complicated shapes are required.

Glass fibre reinforced plastic (GRP) is widely used for:

- Waffle and trough moulds for floor construction.
- Profiled form liners, mainly for wall construction, to provide a patterned finish.
- Special purpose-made forms for complex shapes e.g. circular columns with mushroom heads, etc.

The reuse potential of plastic is, in common with other formwork materials, dependent on its treatment and handling. In general one can expect up to 100 or more uses.

Plastics are particularly susceptible to impact damage and surface scratching, therefore extra care must be taken to avoid hitting the surface. Some plastic forms are specially designed to be removed by compressed air. After striking, the form face should be immediately cleaned to remove all traces of cement dust, using a special cleaning agent, clean soapy water or an oil-soaked cloth. Any hardened cement paste can be removed with a wooden scraper. The use of metal scrapers is not recommended as these damage the surface.

In addition to their high reuse value and their ability to form complex shapes and produce a high-quality finish, they also have the advantage of being low in weight.

Release agents

The correct use of release agents has a significant effect on the quality of the concrete finish and the reuse potential of the formwork. Their main functions are to:

- Provide a quick release from the moulding.
- Prevent adhesion between the form face and concrete surface.
- Aid the production of a high-quality blemish-free concrete finish.
- Enable the maximum reuse of formwork.

There are various types of release agent available. The choice is dependent on the form face material and quality of finish required.

It is therefore essential that care is taken in choosing the correct release agent and that it is compatible with not only the form face but also any other formwork preparation such as barrier paints, fillers, surface retarders and waxes, etc.

The main categories of release agent and surface coatings in general use, together with their main characteristics and suitable form faces, are given in Table 11.3.

Precautions in use

Whatever the category of release agent being used, the best results are obtained with a very thin film. Excessive application causes a poor concrete finish, presents a greater risk of it getting on to the reinforcement and also creates a safety hazard because of the risk of people slipping on horizontal surfaces.

Release agents contain substances that can be harmful to your health. They must therefore always be handled and used with care:

■ Always follow the manufacturer's instructions with regard to use and storage.

■ Avoid contact with skin, eyes or clothing.

■ Avoid breathing in the fumes, particularly when spraying.

■ Keep away from foodstuffs to avoid contamination.

■ Always use the correct PPE:

  – Barrier cream or disposable protective gloves;

  – Respirator when spraying.

■ Do not smoke or use near a source of ignition.

■ Ensure adequate ventilation when used on internal work.

■ Thoroughly wash your hands before eating or smoking and after work with soap and water or an appropriate hand cleanser.

■ In the case of accidental inhalation, swallowing or contact with the eyes, medical advice should be sought immediately.

Loads on formwork

The loads on formwork may be classified as follows:

■ *Dead loads.* These consist of the self-weight of the formwork and the deadweight of the concrete and reinforcing steel.

■ *Imposed loads.* These can be considered as temporary loads and include the impact load of concrete being placed and the live loads of workmen and concreting equipment, in addition to the later storage of materials and forces from the permanent structure.

■ *Environmental loading.* This includes wind loading and snow loading as well as accidental loading.

■ *Hydrostatic pressure.* Caused by the fluid concrete acting against the sides of vertical or steeply sloping formwork. The amount of pressure is dependent on the height of the pour; as the concrete sets the pressure gradually disappears. The maximum hydrostatic pressure

Table 11.3 Release agent

Type	Characteristic	Method of application			Suitable form face					
		Brush	Squeegee	Spray	Sawn timber	Planed timber	Plywood (Unsealed)	Plywood (Sealed)	Metal	Plastic
Neat oil	Encourages the formation of blow holes and therefore not recommended									
Neat oils with surfactant	Contains a surface wetting agent to minimise the formation of blow holes	*		*	*	*	*	*	*	*
Mould cream emulsion	General-purpose release agent particularly suitable for absorbent surfaces	*	*		*	*	*			
Water-soluble emulsion	Causes discoloration and retardation of surface cement and is therefore not recommended for visual concrete	*	*		*	*	*			
Chemical release agent	Causes a chemical reaction at the form face, thus minutely retarding the concrete surface. Recommended for all high-quality work and particularly pretreated surfaces			*	*	*	*	*	*	*
Sealers and coatings, including barrier paints, waxes and other impermeable coatings	Mainly used as pretreatments. Not recommended for use without release agent. They prevent the absorption of release agents into the form face. Wax treatments are particularly useful for filling and sealing small imperfections in the form face	Often factory applied or as directed by the manufacturer			*	*	*	*	*	*
Other specialised release agents including wax emulsion	Formulated for specialised applications such as heated formwork systems and concrete forms	As directed by the manufacturer								
Surface retarders	Although these are not release agents they have the same effect as they retard the setting of the surface concrete. Surface retarders are in fact used to produce an exposed aggregate finish. The surface cement is brushed or washed away after striking the formwork.	*	*	*	*	*	*	*	*	*

Table 11.4 Factors affecting the development of pressure

Factor	Effect on pressure
Density of concrete	Pressure increases in proportion to density
Workability	Pressure increases in proportion to increases in the 'slump' of the mix
Rate of placing	Slow rates of placing enable lower levels of concrete to commence setting before the pour is complete, thus reducing pressure
Method of discharge	Discharge from a height causes a surge or impact loading
Concrete temperature	High temperatures quicken the setting and thus reduce pressure, whereas low temperatures have the opposite effect
Vibration	This makes the concrete flow and increases the pressure up to P_{max}
Height of pour	In theory P_{max} increases with the height of the pour, although in practice during the time taken to fill the form the lower levels will have started setting, and therefore P_{max} over the full height is unlikely to be realized
Dimension of section	Pressure in walls and columns below 500 mm thick is reduced because of the arching or support effect of the aggregate between the formwork sides
Reinforcement detail	Heavy steel reinforcement tends to provide support for the concrete and therefore reduces pressure

(P_{max}) at any point within a form is equal to the amount of concrete above it. Therefore there will be minimum pressure at the top of the form, which will steadily increase to the maximum at the bottom. Factors that affect the development of pressure within formwork are given in Table 11.4

For practical purposes these loads can be considered as wall type loadings, deck type loadings and hydrostatic pressure as illustrated in Figure 11.47.

Basic formwork construction

Formwork plays a very important part in the finished appearance of a concrete structure. Therefore workmanship of a very high standard is required to produce high quality concrete.

When making formwork, its construction depends on the following basic requirements:

■ *Containment.* It must be capable of shaping and supporting the wet concrete until it cures.

■ *Strength.* It must be capable of safely withstanding the dead, imposed and environmental loadings without distortion or danger.

■ *Resistance to leakage.* All joints in the form face must be close fitting or taped to prevent grout leakage.

■ *Accuracy.* The accuracy of the formwork must be consistent with the item being cast. Formwork for foundations is not normally required to be as accurate as formwork for the superstructure.

■ *Ease of handling.* All formwork should be designed and constructed to include facilities for adjustment, levelling, easing and striking without damage to the formwork or concrete.

■ *Finish/reuse potential.* The form face must be capable of consistently imparting the desired concrete finish and at the same time achieving the required number of reuses.

■ *Access for concrete.* The formwork arrangement must provide access for the placing of the concrete.

Formwork for foundations, kickers and columns

Foundations

In firm soil it is often possible to excavate foundations and cast the concrete against the excavated sides. In circumstances where this is not possible, formwork will be required.

Shallow foundations, Figure 11.48, shows typical details of a shallow foundation for a concrete strip, raft or ring beam. It consists of made-up form panels wedged, levelled and strutted in position. Strutting may be from stakes in the level ground or from a sole plate against the bank in deeper excavations. The distance pieces should be removed during pouring as the concrete reaches them. Also shown is a method, which is used to support and locate the column steel reinforcement starter bars. This consists of a simple framework suspended across the form sides and notched around the vertical reinforcement.

Pad bases and pile caps, Figure 11.49, illustrates typical foundation formwork suitable for deeper column pad bases or pile caps. Framed

Wall and column loadings

Minimum pressure at top

Maximum pressure at bottom

Hydrostatic pressure

Deck and beam loadings

▲ **Figure 11.47** Loads on formwork

Distance pieces removed during pour

Steel pin

Edge form

Stakes at 600mm c/c

Sole plate

Struts

Wedge to level edge form

Stakes at 600mm c/c

Sole plate

▲ **Figure 11.48** Shallow foundation formwork

form panels are used for both the sides and the ends. They are wedged to level and then firmly strutted in position, again from either stakes, a sole plate or off the bank. If care is taken during the laying of the blinding concrete to form it to the required plan shape it can be used as a kicker, thus positioning the form and preventing grout leakage. Alternatively, as before, the form sides and base may be lined with polythene sheeting.

The strutting shown in the previous details has been either cut square at both ends and tightened with folding wedges (do not forget to drive a nail through the wedges to prevent them slackening off), or birdsmouthed around a member at one end and cut square at the other. When using the latter method the strut is tightened. Consequently, the form is moved into line or plumb by hammering the square end forward, causing the timbers to bite into each other and create a satisfactory bearing.

Bolt pockets. Often pad bases have to be provided with protruding hold-down bolts or bolt pockets for the later fixing of structural columns or portal frames. A simple framework similar to that used for

▲ **Figure 11.49** Column pad base formwork

locating starter bars can also be used for supporting and locating these hold-down bolts. The bolts are normally required to have a certain amount of tolerance movement in order to permit the later accurate positioning of the structural framework. This can be achieved as shown in Figure 11.50 by the use of tapered polystyrene sleeves around the bolts that can be burnt or melted out after the concrete has been cast.

In situations where several pad bases or pile caps with the same plan size but of different depth are required, the formwork can be constructed to the greatest depth. When these are subsequently used for the shallower pours, battens fixed to the sides can be used to indicate the concrete casting level. A line of nails along the form sides is often used, but this is frowned on by the concreters who may damage their hands on them when trowelling off.

Kickers

A kicker is a small upstand of concrete the same plan size as the finished wall or column but only up to 150 mm in height. Its purpose is to locate the bottom of the form and prevent grout loss. It is an important component in concrete work as far as accuracy is concerned.

Bolt

Locating framework

Tapered polystyrene sleeve

Washer

▲ **Figure 11.50** Locating hold-down bolts

Kickers for walls and columns may be formed together with the base or slab as part of the same pour, or alternatively it may be formed later as a separate operation.

The ideal size for a kicker cast in a separate operation is about 150 mm in height; any more than this and accuracy is lost owing to the difficulty in ensuring that its sides are plumb. Much less than this and the concrete cannot be compacted properly, thus later causing movement at the intersection.

Figure 11.51 shows typical kickers. These are normally held in position by tying them off the reinforcement starter bars with notched battens.

Floating kickers are cast along with the slab and are known as floating kickers or upstands. These are often more difficult to form than separate kickers. Depending on their height, this type of kicker either may be simply supported by previously cast concrete blocks and also tied at intervals back to the edge form with distance pieces, or may be suspended from the edge form with the aid of plywood diaphragm brackets. Concrete blocks may still be used underneath at intervals to reduce any tendency to sag. See Figure 11.52 for details of floating kickers.

In certain situations provision must be made in the kicker design for the later securing of formwork. This normally takes the form of a cast-in form tie. Figure 11.53 illustrates a coil tie being cast in a wall kicker. As well as providing a later securing point it ties the kicker sides

Reinforcement starter bar

Notched batten

Wall kicker

Slotted bolt hole

Off cut nailed to green concrete
to prevent sides bulging

Splayed housing

Column kicker

▲ **Figure 11.51** Wall and column kicker boxes

together, keeping them parallel. Also shown is a loop coil tie cast in a kicker, where a later anchorage point is only required on one side.

For reasons mainly of speed, kickers are sometimes not used, in which case the form may be accurately positioned with the aid of plywood templates, either nailed on to the green concrete, as shown in Figure 11.54 or fixed by cartridge tool.

Columns

A plan view of a typical square column box is shown in Figure 11.55. This is made up in four separate sections; two of these are the same width as the column, whereas the other two are made twice the thickness of the ply over size. Each side is made up so that the backing timbers overhang the edge of the plywood by its thickness, so that the box will be self-aligning when it is cramped up.

In order to prevent grout leakage at corners it is essential that the meeting edges of the plywood are straight and square. As an additional precaution Neoprene or foamed rubber self-adhesive strips can be applied between the mating surfaces.

Large columns will require an additional backing timber placed centrally on each side, or noggins fixed between the backing timbers under each cramp. This is to prevent the ply bending.

slab edge form

distance piece

kicker edge

strut

concrete block

slab edge form

framed plywood diaphragm brackets

strut

framed kicker edge form

▲ **Figure 11.52** Floating kickers

An alternative to the use of proprietary column clamps, which is particularly useful for very large column boxes, is shown in Figure 11.56. This involves the use of yokes made from walings and tie bolts. The spacings of the yokes in common with proprietary clamps will be dependent on the design hydrostatic pressure.

Timber yokes and bolts are also particularly suited for use where L-shaped or rebated columns are being constructed, as illustrated in Figure 11.57.

▶ **Figure 11.53** Kickers with provision for subsequent fixings

▲ **Figure 11.54** Plywood template to position column

▲ **Figure 11.55** Square column box plan

Plumbing columns. The four column sections are held together with cramps or yokes and the whole column box is plumbed up and held in position by an adjustable steel prop on each side. The column must be plumbed in both directions. This can be done with the aid of a plumb bob or straightedge and level.

Figure 11.58 shows both of these methods. The props must be adjusted until the column is plumb. This operation must be repeated on the adjacent edge so that the column is plumb in both directions. When

▲ **Figure 11.56** Use of yokes

Nut and washer
Waling
Backing timber
Plywood formface
Tie bolt
Wedge

▲ **Figure 11.57** L-shaped column

Plywood gusset plate
Yoke bolted at corner

▲ **Figure 11.58** Plumbing a column box

Straight edge with arms and a spirit level
Props adjusted until top and bottom measurements are the same
Plumb-bob suspended from batten fixed to top of box

▲ **Figure 11.59** Proprietary column form

Plywood faced panel
Interlocking key and wedge
Corner connecting angle piece

tightening any prop its opposite one must be slackened otherwise the whole column box will be lifted from the kicker. Immediately after pouring, the column must be rechecked for plumb and carefully adjusted if required, as it may have been knocked during the pouring process.

Proprietary steel-framed panels. Either steel- or plywood-faced panels can be used to form square and rectangular columns. These are connected at their external corners with L-shaped angles that are secured with keys and wedges, as illustrated in Figure 11.59

▶ **Figure 11.60** Plywood mushroom head form

Mushroom heads. Occasionally columns are topped with a splayed or mushroom shaped head. Although they form a distinctive decorative feature, their main purpose is in fact to carry the load of large span floor slabs and transfer these loads on to the columns. The more common method of supporting and transferring loads is to use a system of beams.

Where large numbers of mushroom heads are involved, the use of purpose-made steel or glass reinforced plastic forms would be an economical consideration. Illustrated in Figure 11.60 is a method for forming mushroom heads using timber and plywood. The column is first cast up to the level of the splay. A timber yoke is bolted on to the column shaft at its top to form a platform onto which the head formwork can be erected. This consists of two long plywood sides fixed to shaped plywood formers and bolted together on top of the yoke platform. The other two sides are cut and fixed onto the first two.

Circular columns. These can be constructed using cardboard tube formers, proprietary glass reinforced plastic formers, proprietary steel formers, or made-up timber and plywood forms.

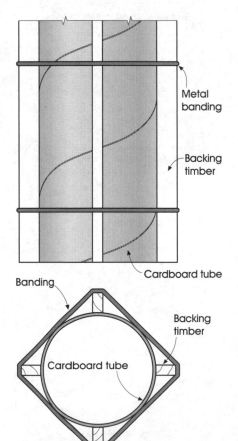

▲ **Figure 11.61** Circular column using a cardboard tube

Figure 11.61 illustrates a section through a cardboard tube former. These require backing timbers and either yokes or metal banding to prevent distortion from the circular during casting. The example shown uses metal banding of the type used in heavy packaging. The bands are tightened around the forms using a special tightening and crimping device. On striking, the bands are simply cut with shears to release the backing timbers. It is customary to leave the cardboard tube in position until occupation in order to provide physical protection for the concrete. Being spirally wound, it is then simply unwound from the column. This method has the advantage of being fairly economical, but the tubes are only suitable for one-off usage.

Illustrated in Figure 11.62 is a method of forming a circular column in timber. It is made up of plywood formers, timber lagging and vertical open spaced timber lagging lined with hardboard or thin ply

step in mating surfaces aids positive location

ply formers at spacings similar to column clamps (e.g. closer at base)

vertical studs

tie bolt at each former

▲ **Figure 11.62** Circular column form

studs in two separate halves which are later bolted together. Adjustable steel props are used for holding and plumbing in the same way as square columns.

Figure 11.63 illustrates an alternative method of constructing circular columns. This uses solid timber yokes and laggings. Where a fair-faced finish is required the laggings may be lined with plywood or hardboard. Four pieces of shaped solid timber are used to form each yoke. Two of these are permanently cleated together to form a semicircle. Subsequently the two halves are bolted together on assembly.

Proprietary or purpose-made circular column forms in either steel or glass reinforced plastic, although initially very expensive, can be a cost-effective consideration where large numbers of columns are required. This is more especially the case in situations where mushroom-headed columns are required.

Figure 11.64 illustrates a section through a column box with shaped ends and a box-out to create a service duct. This is in fact a combination of a square and circular column. It has been made up in two halves using plywood formers, vertical studs and timber laggings lined with either plywood or hardboard. The rectangular service duct has been formed using a box-out that incorporates a false draw. The plywood former and laggings method of column box construction is very versatile; it can be adapted to create any shape of column box.

Bolt

Close timber lagging

Stepped joint

Cleat

Open lagging and ply lining

Solid timber yokes

▲ **Figure 11.63** Circular column using timber yokes

Curve built up using three layers of plywood

Two halves bolted together

Remove this screw before striking

Box out with false drawer to form service duct

▲ **Figure 11.64** Column with shaped end and service duct

375

Formwork for walls

Wall formwork normally consists of a number of standard wall forms bolted together. These are then tied over their backs with either walings or soldiers. The choice of using the form panels vertically with horizontal walings, or of using horizontal panels and vertical soldiers, depends on the nature of the work and the designer. It is general practice for high lifts, where substantial pressures are expected, to use horizontal panels and vertical soldiers.

Figure 11.65 shows typical details of a wall form using standard panels and form ties. These are erected against the previously cast kicker. The tie bolt positions are dependent on the wall height, thickness, method and rate of pour, as these factors greatly affect the development of concrete pressure within the forms.

Adjustable steel props are used at about one metre centres on both sides of the wall to plumb and hold the forms in position. A too steep prop angle should be avoided, as this tends to lift the forms off the kicker. An allowance for shrinkage must be made on long walls. They are often therefore cast in a hit-and-miss pattern so that shrinkage movement can take place before alternate bays are cast. An alternative to adjustable steel props is push-pull props that bolt to the form and the substructure, enabling them to be used on one side of the form only.

Stop ends. Where provision is required in the stop end for the reinforcement starter bars to protrude into the next bay, the detail

▲ **Figure 11.65** Wall forms

Starter bars projecting through stop end

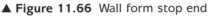
Ply cut to a zigzag around bars for ease of striking

▲ **Figure 11.66** Wall form stop end

Corner

Attached pier

▲ **Figure 11.67** Wall corner

illustrated in Figure 11.66 can be used. This accommodates the starter bars and eases removal after striking the main panels as the two halves of the ply stop end simply slide out from either side.

Corners and piers. Careful formwork detailing of corners and attached piers is required in order to resist the increased pressures at these points. Typical details are shown in Figure 11.67.

Proprietary systems

These are available for wall formwork. Most of these consist of steel framed panels jointed together using keys and wedges. Figure 11.68 illustrates details of one such system. Scaffold tubes are used as walings to tie the panels and align the wall.

Rectangular box-outs, Figure 11.69, shows a box-out suitable for forming door and window openings in wall forms. This is located between the two form faces of the wall and allows the formwork on both sides to be continuous. Keystone-shaped sections are included in each edge of the box. On striking, these sections are simply unscrewed, allowing the corners to be removed separately.

Metal frame panel edge

Plywood form face

Snap tie

Key and wedge joins panels and holds form tie

Scaffold tube for alignment

Clamp-fitting secures scaffold tube and joins panels

▲ **Figure 11.68** Proprietary wall form details

Circular or curved box-outs can be formed using plywood fixed around shaped ribs, as shown in Figure 11.70. To assist removal it has been made in three sections joined using dovetail keys. Again on striking the keys are simply unscrewed to enable the sections to be withdrawn with ease into the hole formed.

Climbing wall forms. Walls are sometimes cast in a number of vertical lifts. This is either to reduce the concrete pressures of a high lift, or to obtain the maximum reuse from a minimum number of form panels. The formwork for this method is known as climbing wall forms.

Figure 11.71 illustrates the basic arrangement for climbing forms. After casting and allowing the first lift to cure sufficiently, the form panels are removed and raised to form the next lift, which is then cast. This process can be repeated until the required wall height is reached.

On fair-faced work the joint line created between lifts may be unsightly. This can be masked by adding a fillet to the panel which forms a

Keystone section unscrewed for striking

Dovetail key

Shaped rib

Screws removed for striking

▲ **Figure 11.69** Rectangular box-out

▲ **Figure 11.70** Circular box-out

vertical section
first lift

these bolts pick
up ties left in
concrete from
first lift

packing

second lift

▲ **Figure 11.71** Climbing wall forms

definite feature joint (see Figure 11.72). This detail is particularly useful on vertical joints in sawn board-marked feature forms.

Single-sided wall forms. In situations where a wall has to be cast against a bank or other structure, the single-sided wall form arrangement shown in Figure 11.73 is used. This has standard form panels and proprietary strong backs. Because of the lack of ties it is essential that each strong back is anchored firmly to the kicker. In the example, this has been achieved by bolting the strong backs to loop coil ties, which were previously cast in the kicker. Where an additional lift is required, loop coil ties could be cast at the top of the wall for subsequent anchorage.

second lift

fillet

recessed feature
masks joint line

first lift

▲ **Figure 11.72** Masking joint line

Plumbing and lining wall forms

Walls are plumbed up and held in position by adjustable steel props on each side. The props at the two ends are adjusted until the wall is plumb using a similar method to that used to plumb columns. Remember when tightening any prop its opposite one must be slackened otherwise the whole wall form will be lifted from its kicker. The top of the wall can then be 'lined in' using the method illustrated in Figure 11.74. Drive nails near the top of the two ends and strain a line between them. Cut three identical pieces of packing, say 18-mm plywood; place one each end under the line. Use the third piece to

▲ Figure 11.73 Single-sided wall form

check the distance between the form face and the line at each support position. Adjust the prop so that the packing piece just fits between the form face and the line.

Curved wall forms

The construction of curved concrete walls may be carried out using panels made up from shaped horizontal ribs, vertical laggings and plywood form face linings, as shown in Figure 11.75. Two different shaped form panels are required – one set to the concrete face internal radius, and the other to the external radius. Twin vertical soldiers and form ties are used to secure the panels where the ribs overlap.

Illustrated in Figure 11.76 is a proprietary system for forming curved wall shapes. This utilises strong backs attached to the plywood form face. The desired internal and external curvatures are formed by adjusting the turn buckles which connect the edge angles that are attached to both ends of the plywood sheets. Adjacent forms may be joined by bolting through these edge angles.

End section

Plumb at both ends

Line strained from end to end on two packers

Use packer to check line at supports

Part plan

▲ **Figure 11.74** Plumbing and lining wall forms

Elevation

Vertical lagging

Ply form face

Form tie

50 mm shaped rib

50 mm x 100 mm twin soldiers

Plan

◀ **Figure 11.75** Constructing a curved concrete wall

Plan

Section

▲ **Figure 11.76** Proprietary system for curved wall shapes

Slabs and beams

Floor and roof slabs

The decking for floor or roof slabs usually consists of plywood, joists and ledgers supported on adjustable steel props, or one of the various proprietary formwork systems available, or a partial combination of both.

Figure 11.77 shows the arrangement of timbers and adjustable steel props for a typical floor slab. Scaffold tubes and fittings have been used for the lacing and diagonal sway bracing. These permit greater prop loadings and prevent sideways movement of the formwork. The recommended positioning for the lacing members is up one-third of the extended inner tube height.

Forkhead fittings are located in the top of the adjustable steel props to provide a secure bearing for the ledgers. Most forkheads are of a size to permit the side-by-side lapping of ledgers. Concentric (through the centre) loading of props is essential if they are to support their design loading. This can be achieved when fork heads are being used for single ledgers by aligning the square head diagonally across the timber and packing out with wedges (see Figure 11.78).

plywood decking

75 mm × 150 mm ledger

75 mm × 100 mm joist

fork head fittings

lacing

adjustable steel props

diagonal bracing

▲ **Figure 11.77** Floor slab decking

The spacing of formwork members will vary depending on the anticipated loads and pressures. As a general guide the following can be used:

■ Props   1.200 m centres
■ Ledgers   1.500 m centres
■ Joists   400 mm centres.

Figure 11.79 shows how a slab may be finished at an open edge. The edge, which sits on top of the decking, is framed up from 50 mm × 75 mm timbers and strutted in position off the 50 mm × 75 mm ribbon with 50 mm × 50 mm struts. The ends of the struts are shown birdsmouthed over the edge framing and ribbon, but in practice these are often cut square.

Drop beams

These are sometimes incorporated at slab edges as shown in Figure 11.80. The beam soffit has been made up using the same method as for slab formwork. The ends of the main slab joists are supported by packing off the beam soffit. The joists supporting the beam soffit have been cantilevered out to enable strutting of the beam side. The cantilevered joists should in turn be strutted back to the adjustable steel

Eccentric loading

Forkhead

Ledger

Concentric loading of forkheads

Forkhead

Wedge

Ledgers

Nail or screw fixing

▲ **Figure 11.78** Forkhead loading

Ribbon and strut

Slab edge

▲ **Figure 11.79** Slab edge form

▲ **Figure 11.80** Drop edge beam

props for support. This cantilevered arrangement is often necessary at upper floor levels where there is no direct bearing available for the adjustable steel props directly below the support position.

Figure 11.81 illustrates formwork details where a beam is incorporated in the middle of a floor slab. Where the beam is deep the beam sides may be held in position by the use of form ties spaced periodically along the beam and positioned towards their top edge.

Careful consideration must be given when determining the layout of plywood soffits where they abut previously cast work or existing structures. This is to avoid any trapping after casting. Figure 11.82 shows how the plywood should be cut around a previously cast column to enable striking. For grout-tight joints against the column, any gap can be filled with a compressible foam or rubber strip.

▲ **Figure 11.81** Drop beam in slab

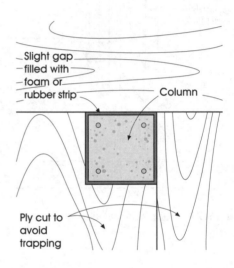

▲ **Figure 11.82** Cutting plywood around columns

▲ **Figure 11.83** Striking strip

Figure 11.83 shows how narrow striking strips may be included along the edges of slabs where they abut beams or walls. On striking, the splayed edge main plywood sheet comes away first, leaving the narrow striking strip adhering to the soffit. This itself is then simply removed by the insertion of a wedge if required. The ends of the joists are also splayed to prevent binding when one end is lowered.

Although in general it is common practice to cast beams and slabs in one operation, beams are sometimes required to span between two supports on their own. Figure 11.84 illustrates the formwork for a beam spanning between two columns. The runners under the beam soffit are not essential, but their use enables wider spacing of the joists and permits easy variation of the beam depth.

Levelling slab formwork and lining edge forms

Datum marks on adjacent walls or columns are used for the initial positioning of slab formwork. Having erected the slab formwork to the correct height, it should be checked for level and adjusted as required. This operation can be carried out with the aid of a quickset level using the following procedure (see Figure 11.85):

1. Set the instrument up and adjust ready for use, adjacent to the slab formwork and at the same level. This may be either on a previously cast section of the same slab or on an adjacent scaffold. Never set up the level on the actual formwork as the instrument will move along with the formwork as it is adjusted.

2. Two assistants are required, one on the decking holding the staff, the other below the decking to adjust the props. It is unlikely that the person below will hear verbal instructions, so prearrange a code system, e.g. one stamp with the foot on the decking to wind up the prop, two stamps to lower it, three stamps for okay.

3. The levels are checked along the lines of each ledger, since it is these that support the joists and decking.

4. Starting at the ledger adjacent to the columns or wall, adjust the props until the top of the decking coincides with the previously established mark. Take a staff reading at either end and midway along this ledger. Each reading should be the same. Record this, as it is the reading required to level the whole deck.

5. Position the staff at one end of the next ledger; adjust the prop until the same reading is achieved. Reposition the staff, first at the other end of the ledger, followed by midway along; adjust props as required to achieve the desired reading. Long ledgers or those formed in two lengths will require additional intermediate checks.

6. Repeat the previous stage for each remaining ledger.

7. After the levelling operation is complete and before loading the formwork, check to ensure that all the prop collars are adjusted tight up to the pin, as the process of adjusting various props can cause the inner tubes of intermediate props to lift up. If the collars were not so adjusted the formwork would sag and distort under load.

Slab edge forms will require lining. This is carried out using a similar method to that used for lining wall forms.

Steel-framed buildings have to be encased to protect against fire and corrosion. Where they are to be encased in concrete it is possible to suspend the formwork from the universal beams using hangers (see Figure 11.86).

Proprietary adjustable floor centres are useful for decking as they reduce not only the amount of timber but also the number of adjustable props required, thus easing the levelling operations and creating a less 'cluttered' under-slab area. Figure 11.87 shows a typical example of the formwork for a floor slab using centres and standard form panels.

▲ **Figure 11.84** Alternative design of beam boxes

Using a quickest level to adjust and level slab formwork

Lining slab edge

▲ **Figure 11.85** Levelling and lining

Drop heads

In situations where early striking of formwork is required, drop heads may be incorporated either at the top of adjustable props or as part of the proprietary support system. Their use enables the decking to be removed three to four days after curing the slab; the support structure remains undisturbed until the concrete has gained sufficient strength to be self-supporting over its full span. Figure 11.88 illustrates drop heads fixed to the top of adjustable props. These are used to support proprietary steel beams and standard form panels. On striking, the drop heads are released using a hammer blow to move the supporting flange universal beam   hanger tie   ply decking

joist

ledger

strut or packing to suit

joist

concrete spacer

she- bolt

twin bearers

◀ **Figure 11.86** Suspended formwork in steel-framed buildings

Make up section   Form panel   Framed beam side

Prop and ledger   Floor centre

Aluminium beam with timber inserts used as ledger

Adjustable hanger used in place of prop

Use of floor centres

Ensure form panels are held in position while striking floor centres

Securing screws slackened to retract and drop floor centres

Striking floor centres

▲ **Figure 11.87** Proprietary floor centres

Before dropping

Form panel

Steel beam

Prop with drop-head bolted on top

Hammer blow on striking to drop beam and form panel

After dropping

Prop head remains to support concrete

Beam and panel can be lifted down from this position

▲ **Figure 11.88** Use of drop heads

to one side. The panels and beams will drop 100 mm or so, and from this position they can easily be handled and lowered.

Table forms

These are popular for the repetitive casting of floor slabs in medium- to high-rise structures. This is because they greatly reduce the time taken to erect, strike and re-erect, due to the fact that these operations can be carried out without any dismantling.

Figure 11.89 illustrates a floor slab, and edge beam detail using table forms as the support system. Where edge beams are concerned, one essential requirement is that the table must be capable of being lowered sufficiently to fit under the edge beam.

Slabs with sloping soffits

These are for use as projecting canopies, etc., and may be formed using framed-up ribs acting as ledgers to support the joists and plywood decking, as shown in Figure 11.90. Cross-braces are fixed to the struts of adjacent ribs to prevent overturning.

Alternatively sloping soffits may be formed as illustrated in Figure 11.91. Here the straight ledgers are raked using differing prop lengths. Wedges cut to the required slope are positioned in the fork head to permit a flat bearing.

▲ **Figure 11.89** Table form edge beam detail

▲ **Figure 11.90** Sloping slab soffit

▲ **Figure 11.91** Alternative sloping soffit detail

Stairs

When setting out formwork for stairs great care is necessary, as the finished result must obviously comply with all the requirements of the Building Regulations, see chapter 5.

In situ formwork for stairs is possibly one of the most expensive operations; therefore wherever possible designers tend to use pre-cast stairs.

Stair soffits can be considered as a sloping slab; thus the supporting formwork will be similar to that of floor slabs. The **step profile** is formed using cut strings and riser boards. **Risers** may be formed from timber, plywood or pressed steel as shown in Figure 11.92. Timber or plywood risers should have a splayed bottom edge to permit the entire tread surface to be trowelled off. The radius along the edges of the pressed steel risers acts as a stiffener and also forms a neat pencil round to the finished step profile.

Two alternative formwork arrangements are illustrated in Figure 11.93. These are both suitable for casting stairs abutting a wall on one side. For stairs that are built between two walls the strutted open edge form

Bottom of riser splayed to enable trowelling and prevent embedding

Radius on edge stiffens risers and forms pencil round

Plywood

Timber

Steel

Soffit

▲ **Figure 11.92** Stairs

32 mm x 75 mm cleats nailed or screwed

Ply soffit

Joists

Strutting off ribbon piece

Close ply string strutted off ribbon piece

Cut wall string

Riser

Cut outer string

32 mm x 75 mm hanger

32 mm x 150 mm wall piece

Joist

Risers fixed to close string using either cleats or an attached cut string

▲ **Figure 11.93** Stair formwork details

can be omitted and substituted by either a wall plate and hangers or a cut wall string. Likewise, free-standing stairs open on both sides can be formed using strutted edge forms on both sides.

Figure 11.94 shows how cut strings may be marked out using a tread/riser template that has been cut to the desired step profile. This is stepped along the string edge the required number of times and marked around with a pencil each time. The outline of steps is best marked out on an abutting wall from the nosing/pitch line. This is an imaginary line that connects the edge of each step. A chalk line is used to mark this on the wall, and the pitch line length (taken from the template) for each step is marked along it. Using a spirit level the horizontal and vertical lines of the step profile can be marked out from these positions. The thickness of the slab concrete is termed the waste (see Figure 11.95).

An inverted carriage piece strutted off the main structure should be fixed down the centre of wide flights to resist the tendency of long slender risers to bulge under concrete pressure (see Figure 11.96). Sectional details showing formwork arrangements for short flights either up to or from landings are illustrated in Figure 11.97.

▲ **Figure 11.94** Marking out a string

◀ **Figure 11.95** Marking stair details on abutting wall

▶ **Figure 11.96** An inverted carriage piece for wide stairs

▲ **Figure 11.97** Formwork for short flights

Pre-cast concrete moulds

The moulds for pre-cast concrete products may be constructed from steel, glass fibre reinforced plastics, or timber and plywood. The first two are mainly the province of pre-casting factories, while timber and plywood are the common materials for use on site.

Construction

Single mound boxes

The basis of mould box construction is a level base or bed on which edge forms can be assembled and cast. The jointing of edge forms is best carried out using bevelled housings. These tighten on bolting, ensuring a grout-tight joint. In addition this detail is less likely to result in damage at the corner of the concrete when striking.

Where not required for wedging purposes, bolt holes for tightening forms are best slotted, since half a turn or so of the nut is all that is required to permit bolt removal. These points are illustrated in Figure 11.98, which shows a simple mould box suitable for casting blocks, slabs or wall panels.

▶ **Figure 11.98** Basic mould box construction

Gang moulds

These are used to cast more than one concrete unit at the same time. Figure 11.99 illustrates a typical gang mould for the pre-casting of tapering fence posts. Placing well-greased steel rods through the distance pieces into the moulds at the required positions permits holes in the posts for bolts or straining wire. These should be removed shortly after the concrete has commenced stiffening.

greased rods to
form holes

wedge

◀ **Figure 11.99** Fence post gang mould

Shaped pre-cast items can be formed by adaptation of the basic mould box design. Figure 11.100 shows how infill pieces are inserted to a rectangular mould box shape to form a weathered pre-cast sill. For speed of erection and striking, long folding wedges have been used on one side of the box instead of the continuous batten.

Figure 11.101 shows how a typical variety of concrete shapes can be formed using the infill technique.

Infill pieces
to suit

Section

Steel rod to
form drip

Long
wedges

Plan

▲ **Figure 11.100** Mould for
pre-cast sill

▲ **Figure 11.101** Infill technique for basic moulds

Cast component slips forward on tilting, breaking bond

▲ **Figure 11.102** Tilting pre-cast bed

In order to facilitate striking, the bed for pre-cast moulds may be pivoted on one edge. After striking the mould box, the bed is tilted to a near vertical position. This causes the cast component to slip a few millimetres on to a restraint batten, and in doing so to break the bond between the concrete and bed face. The concrete component can them be simply lifted off and transported elsewhere for further curing. This tilting bed method is illustrated in Figure 11.102.

Pre-cast stairs

Stairs are an expensive and time consuming in situ casting operation, and so are often specified to be pre-cast. A typical pre-cast stair detail is shown in Figure 11.103. This incorporates rebates at either end, which locate into similar rebates formed at slab landings. Reinforcement loops or cast-in sockets have been provided on the tread surface near the top and bottom of the flight to act as slinging points for crane handling.

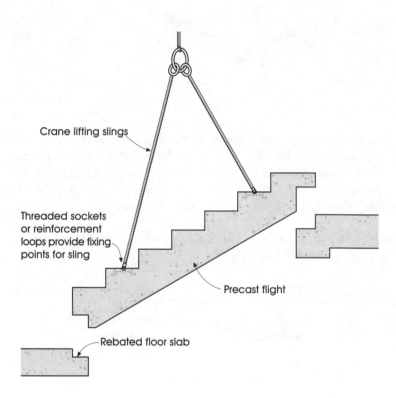

Crane lifting slings

Threaded sockets or reinforcement loops provide fixing points for sling

Precast flight

Rebated floor slab

▶ **Figure 11.103** Dropping in a pre-cast flight of stairs

Three main casting aspects are possible with stair moulds. These are illustrated in Figure 11.104.

■ The inclined casting aspect is really an in situ casting technique applied to a pre-cast unit.

■ Edge casting is useful for fair-faced flights, as all but one edge can have a high quality moulded surface.

■ Inverted flat casting allows the upper surfaces and edges of the flight to be of high-class mould quality, whilst the soffit can be trowelled to an acceptable finish. Inverted flat castings are particularly suitable where tiled nosings or non-slip tread surfaces are to be incorporated, as these may simply be placed in the mould prior to filling with concrete.

Inclined

Flat (inverted)

On edge

◀ **Figure 11.104** Casting aspects for pre-cast stairs

Formwork inspection and safety

At all times during the erection of formwork, and just prior to and during concreting, the formwork supervisor must keep a close eye on the work. This is to ensure that the formwork is to the engineer's specification, capable of producing the required quality (dimensions, tolerances and concrete finish) and, most important of all, safe. The following checklist is useful for this purpose. Owing to the wide range of formwork constructions, the list includes items that may not be relevant to a particular job; therefore it must be amended to suit the work in hand. In addition to the formwork supervisor, similar checks should also be carried out by the general foreman, site engineer, clerk of the works and, where appointed, the false (temporary) work coordinator, one of whom will have the duty of giving permission to load or fill the formwork. During the actual pouring of the concrete it is essential that a fully experienced form worker is on stand-by adjacent to the work to deal with minor adjustments during casting, re-plumbing, levelling or lining after casting and any other contingencies.

Before concreting

- Are all materials suitable and free from distortions, defects, rust and rot, etc.?
- Is the formwork in accordance with the design; in the right position; of the correct dimensions; plumb, level and in line?

■ Are all ties, supports, struts, cramps and braces, correctly positioned, tight and suitably tied or fixed to prevent later movement?

■ Has provision been made for ease of striking without causing damage to the concrete?

■ Have all the stop ends, features, fillets, inserts and box-outs, etc., been correctly located and fixed?

■ Has a release agent been applied to the form face?

■ Have all holes, fixings and joints in the form face been suitably filled or taped?

Checking proprietary equipment

Proprietary equipment is extensively used in all formwork construction. A special check to look out for the following common errors should be made during all site inspections:

■ *Column and beam clamps.* Over-tightening of wedges and bolts, causing crushing of backing timbers and subsequent loss of strength; no support under clamps (nails or blocks) to prevent them from slipping; projecting end of clamps not painted white or bright to make them obvious; loose or incorrect location of wedges.

■ *Floor centres.* Incorrect location of, or failure to tighten, locking nuts or wedges; bearing plates inadequately seated; centres used at their minimum span (no provision to retract for striking).

■ *Form ties.* Too small or incorrectly positioned washer plates; over-tightening of bolts; unequal size of soldiers or walings; damaged or substandard threads (Figure 11.105).

■ *Adjustable steel props.* Use of visibly bent, damaged or corroded props; use of nails, bolts or steel reinforcement in place of the high tensile steel pins supplied by the manufacturers; missing or

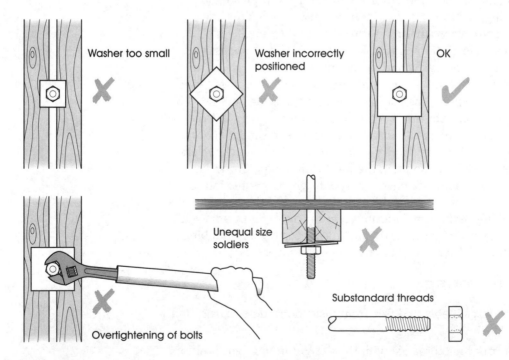

▲ **Figure 11.105** Checking form ties

Wedges

Fork head

Concentric loading

Eccentric loading

Fork head concentrically loaded

Head plate flat and welds intact

Jube not bent, dented or pitted

Ensure correct high tensile pin not make shift nail or reinforcement bar

Prop plumb

Base plate flat, welds intact and centred on sole plate if required

▲ **Figure 11.106** Checking adjustable steel props

inadequate bracing and lacing; props out of plumb (check with a spirit level); props eccentrically loaded; inadequate bearing at either top or bottom (Figure 11.106).

During concreting

Look out, listen for and bring to the attention of the engineer any of the following:

■ Formwork straining noises (ties snapping, timber creaking, etc.).

■ Any movement or distortion in the formwork.

■ Any joints opening or grout leakage.

Each of these factors may be an indication of a possible formwork blow-out or collapse. Common failures commence with grout leakage followed by distortion and bulging in the formwork. In cases where a form tie fails its neighbouring ties become overloaded and rapid progressive failure can result. The immediate action is to stop the pouring and vibration immediately, evacuate personnel from potentially dangerous situations, and then take remedial action to relieve and control the excessive pressures.

Where appropriate, check that distance pieces have been removed and fixing blocks, etc., inserted.

Note: There are other items, which must be checked but are not usually the responsibility of the formwork carpenter. These include: safe access, steel reinforcement, key to previously poured surface, availability of concrete and equipment, method and rate of fill, curing, etc.

Striking

The time between the pouring of the concrete and the striking of the formwork varies widely. This depends on the following four factors:

- The type of formwork, e.g. wall, column, slab or beam.
- The concrete used and the type of surface finish required.
- The weather conditions and exposure of the site.
- The requirements of the specification.

Minimum striking times are shown in Table 11.5 for ordinary Portland cement (OPC).

Table 11.5 Minimum striking times

Location of formwork	Temperature		
	7°C	16°C and over	Formula for t between 0°C and 25°C
Vertical surfaces			
Walls, columns and beam sides	18 hrs	12 hrs	$300 \div (t + 10)$ hrs
Horizontal surfaces			
Slab soffits	6 days	4 days	$100 \div (t + 10)$ days
Slab support structure	15 days	10 days	$250 \div (t + 10)$ days
Beam soffits	15 days	10 days	$250 \div (t + 10)$ days
Beam support structure	21 days	14 days	$360 \div (t + 10)$ days

Example

To determine the minimum striking time for slab formwork, assuming a temperature 10°C during curing.

Slab soffit

$$\text{Minimum time} = \frac{100}{t + 10}$$

$$= \frac{100}{10 + 10} = 5 \text{ days}$$

Support structure
Minimum time $= \dfrac{250}{t + 10}$

$= \dfrac{250}{10 + 10} = 12.5$ days

- No formwork must be removed until the concrete has cured sufficiently for it to be self-supporting and capable of carrying any loads imposed upon it.

- Vertical surfaces such as walls, columns and beam sides etc. can, in very good weather, be struck as little as nine hours after pouring.

- Horizontal surfaces normally require a much longer time, and this can be up to twenty-eight days in poor winter weather. It is usual after striking horizontal formwork to re-prop the concrete in strategic positions for up to a further fourteen days.

- The final decision to strike must always rest with the site engineer, who has knowledge and experience of what is involved.

- The safe time to strike the formwork can be determined by tests on cubes made from the same mix as the concrete and stored under similar conditions.

Safety

Striking is an important operation, the procedure for which must be carefully considered to avoid any safety risk to personnel, stress or damage to the completed building, and damage to the formwork itself.

In general, the procedure to adopt is the reverse of that used for erection. Observe the following precautions:

- Always wear appropriate PPE personal protective equipment (safety boots, helmet, gloves, etc.).

- Never stand or allow anyone else to stand close to the striking area.

- Use hardwood wedges to ease the forms away from the concrete surface (the use of nail bars results in both form and concrete damage).

- Always release (ease) loaded member gradually. For example, props for floor slabs must be released in small stages (half a turn of the handle) starting from the centre and working out, then repeating the operation.

- Always control the striking by removing a section at a time.

- Never allow the formwork to fall (crash striking).

- Always clean, de-nail, repair and store under cover, ready for reuse, all formwork material. In addition all props, cramps and bolts, etc. should be cleaned, lubricated and coated with a rust inhibitor.

Timbering to excavations

Timbering is the temporary support given to the sides of excavations, in order to prevent soil collapse whilst construction work is taking place. The term 'timbering' is normally applied irrespective of the materials used.

The sides of excavations are prone to collapse in on themselves unless they are either battered back or supported with timbering. The angle of battering is dependent on the type of subsoil, the presence of ground water and the time of year/weather conditions. An angle of up to 45° is suitable in most circumstances, see Figure 11.107.

▲ **Figure 11.107** Battering or timbering to foundation trenches

Open timbering

Open timbering is where the sides of the excavation are supported at intervals and are not completely covered. This method is normally use for excavations in firm and moderate ground.

Figure 11.108 shows the stages undertaken when erecting open timbering to the sides of a foundation trench. As the trench is mechanically excavated, pairs of vertical poling boards are positioned in the trench and temporarily held tight to the sides with struts.

Horizontal pairs of walings are introduced once 2 to 3 m has been excavated. These are held in position against the poling boards by struts at between 1.2 and 1.8 m centres. Wedges may be required between the poling boards and walings to ensure each board is held tight up to the earth.

Finally on completion the temporary struts to the individual poling boards can be removed, to give working space.

Closed timbering

Closed timbering used in loose, soft or waterlogged ground, is where the sides of the excavation are completely covered.

Poling boards

Wedges as required

Poling boards

Temporary strut

Strut

Temporary support stage

Walings

Final arrangement of timbering

◀ **Figure 11.108** Open timbering

Figure 11.109 shows the stages undertaken, when erecting close timbering to the sides of a foundation trench. In this case poling boards (either timber or metal) are driven into the ground before any excavation is undertaken.

As the ground is excavated between the poling boards, temporary short lengths of walings and struts are inserted for support. As before when a sufficient length has been excavated, continuous pairs of walings and struts can be inserted.

Close poling boards

Strutlip

Wedges as required

Temporary walings and struts

Puncheon

Strut

Poling boards driven in before excavation

Trench excavated temporary walings and struts inserted for support

Walings

Final arrangement of timbering

▲ **Figure 11.109** Close timbering

Vertical puncheons are used at intervals between the walings to stop them slipping. The struts may be cleated to the walings for additional security using short lengths of batten, called strutlips. Again wedges may be required between the poling boards and walings to ensure each board is held tight up to the earth.

Finally on completion the temporary walings and struts to the poling boards can be removed, to give working space.

Hoardings and fencing

Hoardings

Hoardings are, in effect, screens, which shield areas such as wastelands, allotments, demolition sites and building sites from public view for various reasons, for example, protection, safety, security and aesthetics.

Figure 11.110 shows typical hoarding details. They are often constructed using panels which are framed up in sections from 50 mm × 100 mm timbers, and sheeted with 18-mm WBP plywood. These panels are secured at their base by 50 mm × 50 mm stakes, which are driven approximately 600 mm into the ground. Braces are fixed to the tops of the panels to hold them plumb. These braces are also fixed at their bottom end to 50 mm × 50 mm stakes. Where the hoarding is adjacent to a scaffold, the stakes and braces can be omitted and the panels fixed back to the scaffold using hooks and bolts (providing the additional wind loading has been allowed for in the scaffold design).

Viewing screens are sometimes incorporated in the hoarding to give the public a view of the building works that are going on behind it. Where there is the possibility of a large amount of public interest in the building, which could result in congestion on the pavement, a short set-back in the hoarding could be formed to create a viewing area. If personnel access is required through the hoarding, a wicket gate in the panel as shown can provide this. For vehicular access, several panels could be hinged to form a large pair of gates. Temporary walkways for public protection may be required where a hoarding partially blocks a pavement (Figure 11.110 detail).

Boundary fences

Timber fences can be classified into two main types. These are:
- *Open-boarded fences.* These define the boundary by providing a visual barrier.
- *Close-boarded fences.* These, as well as defining the boundary, also provide a certain amount of privacy and security.

Open-boarded fences

These are usually constructed from loose fencing materials in a variety of designs.

▶ **Figure 11.110** Hoarding and details for hoarding with temporary footpath

Ranch-style fence, see Figure 11.111, consists of 25 mm × 150 mm boards, screwed to the face of the 75 mm × 75 mm posts. For improved durability the boards can be partly recessed into the posts. When it is necessary to lengthen the boards they can be butted midway over a post. These joints should be staggered so that adjacent boards are not butted on the same post. The matching gate is made in a similar way, using two stiles and a brace.

Picket or palisade fence, Figure 11.112, shows a simple method which consists of 75 mm × 75 mm posts, 50 mm × 75 mm weathered edge rails and 25 mm × 75 mm shaped top pickets or paling. The rails can either be mortised and pinned into the posts or fixed using a metal bracket. Again a matching gate can be made in a similar way, using two rails and a brace, with the pickets nailed on the face.

Stockade or interlap fence, see Figure 11.113. The stockade is a taller version of the picket fence. For increased privacy boards can be fixed alternately to both sides of the rails to form an interlap fence.

▲ **Figure 11.111** Ranch-style fence

▲ **Figure 11.112** Picket fence

▲ **Figure 11.113** Stockade or interlap fence

The gap between the boards is a matter of personnel preference; narrow gaps are used for privacy and wider ones simply to define the boundary with a decorative visual barrier. This type of fence is ideal for use in exposed situations as the gaps between the boards allow strong winds to pass through, reducing the amount of wind pressure exerted on the fence.

Close-boarded fences

These may be constructed from either loose fencing materials such as featheredge boarding or using prefabricated fencing panels.

Featheredge fence. Figure 11.114 shows a close-boarded featheredge fence. This is made up from 100 mm × 100 mm or 100 mm × 125 mm posts, 75 mm triangular arris rails and 19 mm × 100 mm featheredge boarding. Fences up to 1.2 m in height require two arris rails, each position about 150 mm above and below the intended ends of the featheredge boards. For taller fences up to 1.8m in height, an additional centre rail should be fitted. The ends of the arris rails are best tenoned into the posts and secured in position using dowels; as an alternative they can be fitted using metal brackets. The capping and gravel board are often omitted on cheaper work, but this is a false economy as their purpose is to prevent moisture absorption into the end grain of the feather-edge boarding, which if allowed to take place would quickly result in decay. Each piece of featheredge should overlap by about 20 mm and be fixed by one nail through the thick edge at each rail. Care must be taken to ensure the nails do not go through the thin edge of the board below where they overlap, as it will split when subjected to moisture movement. The entire run of the posts and rails should be erected and secured before boarding. Gates for close-boarded fences are usually made up in the same way as ledged and braced doors.

Arris rail fixed to post with mortice and tenon or metal bracket

Capping

Post cap

Capping batten

Feather-edge boarding

Arris rail

Use a spacer to get boards even

Arris rail

Boards fixed with one nail per board at each arris rail position

Gravel board cleated or housed into post

Stake to fix centre of gravel board

▲ **Figure 11.114** Feather-edge close-boarded fence

Prefabricated fence panels. Standard fence panels are 1.8 m wide and range in height from 600 mm to 1.8 m, increasing in 300 mm intervals. Most fence panel are made from horizontal overlapping thin strips of timber, often waney edged, which are sandwiched between a sawn timber frame. These panels are sold as lap or waney lap panels, see Figure 11.115. Again the use of capping and gravel boards is

Panel clipped or nailed to post

Post cap

Post

Panel capping

Waney edge strip

Edge framing

Gravel board

Metal fence post socket and spike

◀ **Figure 11.115** Prefabricated panel fence

recommended for extended fence life. After positioning and securing the end post, the first panel is fixed to it using nails through the edge frame or metal clips. The entire fence is constructed by erecting posts and panels alternately, Matching gates are available from many suppliers.

Fence posts

Timber posts are traditionally used for most types of fence either 75 mm or 100 mm square. Hardwood (oak) posts are more durable, but preservative treated softwood posts are cheaper. All timber posts should be either weathered at the top or capped, to shed rainwater thus prolonging their life, see Figure 11.116.

Concrete posts are made in a variety of styles from pre-cast concrete. Mortised posts are used with arris rails and grooved posts for panel fences as illustrated in Figure 11.117. They are considered to be stronger than timber posts and are also rot free. Concrete gravel boards are available for use with grooved posts.

Erection of fence posts

The type of fence will determine whether all of the posts are erected and secured first, or they are put up one at a time.

The first task is to establish the line of the fence. Drive pegs into the ground at either end along the boundary and stretch a builder's line between them to align the row of posts. These should be spaced at between 1.8 m and 3 m centres, depending on the size and type of fence. Where the fence runs up to a property, the first post can be fixed to the wall using expanding masonry bolts.

Fence posts may be secured in the ground using either hardcore and concrete or proprietary metal spiked sockets.

Hardcore and concrete footings. Use the following sequence of operations, see Figure 11.118:
1. Prepare hole for the post. The depth of the post into the ground should be 600 mm for fences up to 1.2 m in height and 750 mm for fences up to 1.8 m.
2. Ram some hardcore (broken brick or stones) in the bottom of each hole, to give a firm base for the post ends and provide drainage.
3. Plumb the post with a sprit level and temporarily brace to hold in position.
4. Ram more hardcore around the post up to a depth of 300 mm
5. Fill around the post with concrete, tamping with a batten to ensure compaction.
6. Finish the top of the concrete just above ground level and smooth it off so that it slopes away from the post, to shed rainwater and reduce the risk of rotting in timber posts.
7. Allow the concrete to set before removing the temporary braces.

Proprietary metal spiked sockets. These do not require any holes digging; they are simply driven into firm ground. Use 600 mm spikes

▲ **Figure 11.116** Timber fence posts

▲ **Figure 11.117** Concrete fence posts

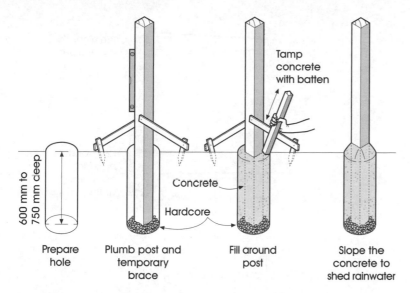

600 mm to 750 mm ceep

Concrete

Hardcore

Tamp concrete with batten

Prepare hole

Plumb post and temporary brace

Fill around post

Slope the concrete to shed rainwater

◀ **Figure 11.118** Concreting in a fence post

for fences up to 1.2 m high and 750 mm ones for those up to 1.8 m. Bolt-down sockets are also available for use where posts are to be erected on to previously concreted surfaces see Figure 11.119.

The following sequence of operations is used to insert metal fence post spikes:

1. Insert an off-cut of post or a proprietary driving block into the spike socket.
2. Drive the spike partway into the ground using a sledgehammer.
3. Check the socket for plumb using a sprit level.
4. Continue driving the spike into the ground until only the socket is visible. Re-check for plumb at intervals to ensure you are still driving the spike in vertically.
5. Insert post into socket and secure it by either screwing or tightening the bolt, depending on the make and type of spike.

Bolt down socket

Spiked socket

Lug for fixing gravel board

Part drive spike and check plumb

Complete driving until only socket is visible

Insert post and either screw or bolt

▲ **Figure 11.119** Metal fence post socket

Automatic gate latch
for small gates

Tee hinge for
general use

Strap hinges

Hook and
band for
heavier
gates

Ring latch
or
thumb latch
for taller gates

Rim lock and
tower bolt
or
padlock, hasp
and staple for
security

Coil spring to provide
a self closing action

▲ **Figure 11.120** Typical gate ironmongery

Weathered
capping

Door
frame

Ledged
and
braced
gate

▲ **Figure 11.121** Weathered
capping used for external door
frames

Hanging gates

Most gates are simply hung between two posts, using strap hinges, see
Figure 11.120. Low gates are held closed by an automatic gate latch;
taller gates are fitted with a thumb or ring latch. A coil spring may be
added to provide a self-closing action. Where security is required, tower
bolts or a padlock, hasp and stable may be fitted to the inside of the
gate. Alternatively a rim or mortise lock may be used.

Door frames are sometimes used for side entry gates, these should
be fitted with a weathered capping for protection as illustrated in
Figure 11.121.

Preserving fences

All timber that is used for fencing must be preservative-treated. This can
either be pressure treated or done by steeping the fencing material in
preservative. Where the fence is to have a painted finish, for example,
small picket or ranch-type fences, the timber should be treated with a

preservative that is capable of receiving a paint finish. The preservative treatment of the posts is particularly important, as untreated or ineffectively treated timber that is in contact with the ground will rot in a very short space of time. It was often for this reason that traditionally durable oak posts were used. Coating the bottom part of the post with bitumen, thereby forming a weatherproof skin, can prolong their lifespan.

In order to keep the timber posts clear of the ground, or where the posts of existing fences have rotted through at ground level, they may be repaired by bolting to a short concrete spur post that has been concreted behind the timber post (see Figure 11.122).

▲ **Figure 11.122** Concrete spur

Chapter Twelve

Repairs and maintenance

This chapter covers the work of both the site carpenter and bench joiner. It is concerned with principles of undertaking repairs and maintenance to buildings. It includes the following:

- ¶ planned and corrective maintenance
- ¶ reasons for maintenance
- ¶ repairs and maintenance of timber components and associated fittings
- ¶ re-glazing and painting as part of a maintenance programme
- ¶ multi-skill repairs to brickwork, plasterwork and tiling as part of a maintenance programme.

Maintenance is taken to mean the keeping, holding, sustaining, or preserving of a building and its services to an acceptable standard. This may take one of two forms: planned maintenance and unplanned maintenance or repairs.

Maintenance and repair routines

There are a variety of reasons why products and processes fail. The impact of many of these failures can be reduced or contained by maintenance routines.

Planned or routine maintenance

This is a definite programme of work aimed at reducing to a minimum the need for often costly unplanned repair work. It includes:

- The annual inspection and servicing of general plumbing, heating equipment, electrical and other services.
- The periodic inspection and cleaning out of gutters, gullies, rain water pipes and airbricks, etc.
- The periodic redecoration, both internally and externally.
- The routine general inspection/observation of the building fabric and moving parts.

Also included under this heading, is what is known as preventative maintenance. Basically this is any work carried out as a result of any of the previous inspections in anticipation of a failure, for example, the early replacement of an item, on the assumption that minor faults almost certainly lead onto bigger and more costly faults unless preventative work is carried out.

Unplanned emergency or corrective maintenance

This is work that is left until the efficiency of the element or service falls well below the acceptable level or even fails altogether. This is the most expensive form of maintenance, making inefficient use of both labour and materials and often creating serious health/safety risks. Yet it is the type most often carried out. This is because the allocation of money to enable maintenance work to be planned is often given low priority.

Causes of failure in products, processes and materials

Apart from the natural ageing process of all buildings during their anticipated life (however well maintained), early deterioration of buildings can be attributed directly to one or often a combination of the following agents:

- design and construction factors;
- chemical attack: effect on surface finish of chemical pollution in the atmosphere;
- weathering and damage: the effect of frost, rain, snow, sun and wind;
- dampness: the effect of wet when allowed to penetrate into the building;
- biological attack: fungal decay and insect attack.

Design and construction factors

Faulty design and construction methods can lead to rapid deterioration of a building. In fact approximately one third of all maintenance/repair work could be avoided if sufficient care was taken at the design and construction stages.

Faulty design

This results from inadequate knowledge or attention to detail on the part of the architect or designer leading to, for example, poor specification of materials/components, structural movement, moisture penetration, biological attack and the inefficient operation of the building services.

Faulty construction

Inadequate supervision during the construction process can result in poor workmanship, the use of inferior materials and the lack of attention to details/specifications. These can all lead to similar problems as those stated for faulty design, resulting in subsequent corrective measures and expense for the building owner.

Dampness

In buildings this is the biggest single source of trouble. It causes the rapid deterioration of most building materials, can assist chemical attack and creates conditions which are favourable for biological attack. Dampness can arise from three main sources: rain penetration, rising damp and condensation. In addition, leaking plumbing and heating systems and spillage of water in use are also significant causes of dampness.

Rain penetration

This is rain penetrating the external structure either through the walls, windows or the roof and appearing on the inside of the building as damp patches. After periods of heavy rain these patches will tend to spread and then dry out during prolonged periods of dry weather. They will, however, never completely disappear, as a moisture stain and in some cases even efflorescence (crystallised mineral salts) will be left on the surface.

Mould growth (fungi resulting in dark-green or black patchy spots) may occur in damp areas particularly behind furniture, in corners and other poorly ventilated locations. The main causes of rain penetration are shown in Figure 12.1. It can be seen that penetration takes place through gaps, cracks, holes and joints in, around or between components and elements.

Roofs

Loose or missing tiles or slates, including the hip and ridge capping tiles, will allow rainwater to run down rafters, causing damp patches on the ceilings and tops of walls. These patches may appear some distance away from the defective area as the water spreads along timbers and across the ceiling, etc. This dampness will also saturate any thermal insulation material making it ineffective. If left unrepaired, saturation of the roof timbers will occur leading to fungal decay in due course. Another major area of penetration is around the chimneystack and other roof-to-wall junctions; this may be due to cracked chimney pots, cracked or deteriorating flaunchings (the sloping mortar into which the pots are set), or corroded or pitted metal flashings (these cover the joint between the stack or wall and the roof), which may be cracked or deteriorated. Poor pointing to the stack can also be a cause of penetration. Any of these defects can cause large patches of damp on an internal wall.

Walls

Clearly rainwater travels downwards and when assisted by high winds it will travel sideways through gaps. But depending on the nature of the material, it can often move unassisted both sideways or upwards because of capillary attraction. **Capillarity** is the phenomenon whereby water can travel against the force of gravity in fine spaces or

▲ **Figure 12.1** Rain penetration

between two surfaces which are close together; the smaller the space the greater the attraction.

There are two main conditions that promote capillarity in the external envelope. The fine cellular structure of some materials provides the interconnecting pores through which water can travel. Also the fine joints between components, for example, wall to door or window frame, mortar joints between brickwork, close joints between overlapping components. The risk of capillarity is reduced or avoided by either:

- Physically separating the inside and outside surface by introducing a gap (e.g. cavity wall construction).
- Introducing an impervious (waterproof) barrier between components (e.g. mastic pointing, DPCs, DPMs, moisture barriers and flashing, etc.).

Over time, the water resistance of brick/stonework and their mortar joints will deteriorate. This deterioration can be accelerated by the action of frost. Rainwater may accumulate below the surface and freeze. Ice expands causing the brickwork/stonework and their mortar joints to spall (crumble away). The wall then offers little resistance to the weather and should be replaced. This entails either:

- Cutting the surface of the spalled components back and replacing with matching thin components (half bricks) and finally re-pointing the whole wall.
- Or the entire wall may be 'hacked off' (cut back to remove spalling) and covered with one of the standard wall finishes, cement rendering, rough cast, pebble dash, Tyrolean or silicone-nylon fibre.

Cracks in cement rendering and other wall finishes can be caused by shrinkage on drying, building movement or chemical attack. Once opened up, deterioration is accelerated by frost action. Small cracks may be enlarged and filled with cement slurry. Large areas that may have 'blown' (come away) from the surface will require hacking back to a sound (or firmly adhering) work surface and replaced. With cavity wall construction, rainwater that does penetrate the outer leaf should simply run down inside the cavity and not reach the internal leaf. The vertical mortar joints of the outer leaf are sometimes raked out at intervals along the bottom of the cavity to provide weep holes through which the water can escape.

However, when the cavity is bridged by a porous material (e.g. the collection of mortar droppings on the wall ties during construction), the water will reach the inner leaf causing small isolated damp patches on the internal wall surface. The remedy for this fault is to remove one or two bricks of the outer leaf near the suspected bridge and either clean out or replace the tie as necessary.

Dampness around door and window frames is likely to be caused by wind-assisted rain entering the joint between the wall and frame by the action of capillarity or by a defective vertical DPC used around openings in cavity walls, where the inner and outer leaf join. Exterior mastic can be used to seal the joints but where DPCs are defective they will require cutting out and replacing. A check should be made at the sill level of frames. Cracked sills allow water to penetrate and therefore should be filled. The drip groove on the underside of the sill should be cleaned out as it often collects dirt/dust and is filled by repeated painting. The purpose of the drip groove is to break the under surface of the sill making the water drip off at this point and not run back underneath into the building.

Blocked or cracked gutters and down-pipes, dripping outside taps and constantly running overflows can cause an excessive concentration of water in one place which will be almost permanently damp. This will result in an accelerated deterioration of the wall and subsequent internal damp patches, etc. Repairing or replacing the defective component can easily rectify the immediate fault. But if left unattended the resulting damage to the building structure has most serious and costly implications.

Rising damp

This is normally moisture from below ground level rising and spreading up walls and through floors by capillarity. This most often occurs in older buildings. Many of these were built without DPCs and DPMs, or, where they were incorporated have broken down possibly with age (e.g. slate, a one time popular DPC material which cracks with building movement, thus allowing capillarity). The visual result on the walls is a band of dampness and staining spreading up from the skirting level, wallpaper peeling from the surface and signs of efflorescence. The skirting, joists and floorboards adjacent to the missing or failed DPC are almost certain to be subject to fungal attack. Solid floors may be almost permanently damp causing considerable damage to floor coverings and adjacent timber/furniture, etc. Rising damp can still occur in buildings that have been equipped with DPCs and DPMs.

One of the main reasons for this is the bridging of DPCs; in the case of cavity walls, builders' mortar droppings or rubble may have collected at the bottom of the cavity, allowing moisture to rise above the DPC level; or earth in a flower bed being too high above the DPC. DPCs are normally located at least two courses of brickwork (150 mm) above the adjacent ground level. This is because even very heavy rain is unlikely to bounce up and splash the walls much more than 100 mm from the surrounding surface. Thus the splashed rainwater is still prevented from rising above the DPC. Where the surrounding surface is later raised these splashes might bypass the DPC and result in rising damp. Weak porous rendering, which has been continued over the DPC is another means by which the DPC may be bypassed (see Figure 12.2).

▲ **Figure 12.2** Rising damp

In solid floors with a DPM, rising damp can only occur if this is defective. For example, it may have been penetrated by jagged hardcore during the pouring of the concrete or may have been inadequately lapped (permitting capillarity between the lapped joint) or finally it may not have been linked in with the DPC in the surrounding walls (allowing moisture to bypass at this point).

The remedy to rising damp faults will of course vary. Bridged or bypassed DPCs can be rectified by simply removing the cause, for example, lowering the ground level or removing mortar and rubble from the cavity, etc. Where the DPC itself is faulty or missing altogether, one can be inserted by either cutting out a few bricks at a time to allow the positioning of a new DPC, or sawing away the mortar joint a section at a time and inserting a new one.

Alternatively, liquid silicone may be injected near the bottom of the wall. This soaks into the lower courses, which then acts as a moisture barrier preventing capillarity.

Localised faults in DPMs can be remedied by cutting out a section of the floor larger than the damp patch, down to the DPM, taking care not to cut through it. This should reveal the holed or badly lapped portion, which can be repaired with a self-adhesive DPM. An alternative method, which can also be used in floors without any DPM, is to cover the existing concrete floor with a liquid bituminous membrane or a sheet of heavy-duty polythene sheeting before laying a new floor finish, although, to be effective it should be joined into the DPC.

Condensation

The results of this form of dampness are often mistakenly attributed to rain penetration or rising damp, as they can all cause damp patches, staining, mould growth, peeling wallpaper, efflorescence, the fungal attack of timber and generally damp, unhealthy living conditions. The water or moisture for most condensation actually comes from within the building. People breathing, kettles boiling, food cooking, clothes washing and drying, bath water running, etc., are all processes which add more moisture to the air in the form of vapour which can lead to **surface condensation**.

Air is always capable of holding a certain amount of water vapour. The warmer the air, the more vapour it can hold but when air cools, the excess vapour will revert to water. This process is known as condensation. Thus whenever warm moist air meets a cool surface condensation will occur (see Figure 12.3). This can only be controlled effectively by achieving a proper balance between heating, ventilation and insulation. The building should be kept well heated but windows should be opened or mechanical ventilators used especially in kitchens and bathrooms to allow the vapour-laden air to escape outside and not spread through the building. External walls need thermally insulating to remove their cold surfaces. Both cavity wall insulation and lining the walls with a thin polystyrene veneer help a great deal. Double glazed windows also help reduce condensation by preventing the warm moist air coming into direct contact with the cold outside pane of glass.

In addition to this surface condensation, there is another condensation problem that occurs when wall surfaces are warm. This is known as

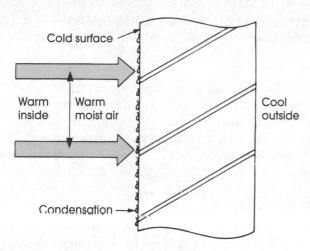

▲ **Figure 12.3** Surface condensation

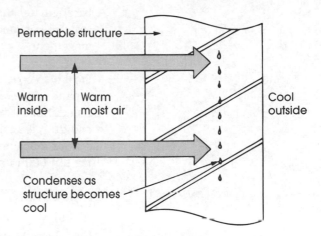

▲ **Figure 12.4** Interstitial condensation

interstitial or internal condensation. This is illustrated in Figure 12.4. The warm moist air passing into the permeable structure until it cools, at which point it condenses leading to the same problems associated with penetrated and rising dampness.

Interstitial condensation can be dealt with:

■ by the use of a **vapour barrier** (this prevents the passage of water vapour) on the warm inside of the wall, e.g. a polythene sheet or foil-backed plasterboard;

■ by allowing this water vapour to pass through the structure into a cavity where it can be dispersed by ventilation.

Movement

The visual effects of movement in buildings may apparently be of a minor nature, e.g. windows and doors that jamb or bind in their frames; fine cracks externally along mortar joints and rendering; fine cracks internally in plastered walls and along the ceiling line etc. They can however be the first signs of serious structural weakness. Movement in buildings takes two main forms these being: ground movement and movement of materials.

Ground movement

Any movement in the ground will cause settlement in the building. When it is slight and spread evenly over the building it may be acceptable, although when more than slight or differential (more in one area than another), it can have serious consequences for the building's foundations and load-bearing members, requiring expensive temporary support (shoring) and subsequently, permanent underpinning (new foundations constructed under existing ones).

Ground movement is caused mainly by its expansion and shrinkage near the surface, owing to wet and dry conditions. Compact granular ground suffers little movement, whereas clay ground is at high risk. Tree roots cause ground shrinkage owing to the considerable amounts of water they extract from it (Figure 12.5). Tree roots can extend out in all directions from its

Tree roots cause shrinkage leading to foundation settlement

▲ **Figure 12.5** Ground movement

417

base, greater than its height. In addition, overloading of the structure beyond its original design load can also result in ground movement.

Frost also causes ground movement. Water in the ground on freezing expands. Where this is allowed to expand on the undersides of foundations it has a tendency to lift the building (known as frost heave) and drop it again on thawing. This repeated action often results in serious cracking. Freezing of ground water is limited in the UK to about the top 600 mm in soil.

Movement in materials

All building materials will move to some extent owing to one or more of the following reasons: temperature changes, moisture-content changes and chemical changes. Provided the building is designed and constructed to accommodate these movements or steps are taken to prevent them, they should not lead to serious defects.

- *Temperature changes* cause expansion on heating and shrinkage on cooling; particularly affected are metals and plastics, although concrete, stonework, brickwork and timber can be affected also.

- *Moisture changes* cause expansion when wetted and shrinkage on drying. This is known as moisture movement. The greatest amount of moisture movement takes place in timber, which should be painted or treated to seal its surface. Moisture movement can also affect brickwork, cement rendering and concrete. Rapid drying of wetted brickwork in the hot sun can result in cracks, particularly around window and door openings.

Chemical and biological attack

Chemical

This consists of the corrosion of metals and the sulphate attack of cement. Corrosion causes metals to expand and lose strength. Corrosion of steel beams can lift brickwork causing cracks in the mortar joint. Corroded wall ties may cause bulges in cavity brickwork. The sulphate attack of cement occurs either in the ground or through products of combustion in chimneys. The sulphate mixes with water and causes cement to expand. Sulphate-resisting Portland cement (SRPC) should be used in conditions where high levels of sulphate are expected.

Smoke containing chemicals is given off into the atmosphere as a result of many manufacturing processes. This mixes with water vapour and rainwater to form dilute or weak acid solutions. These solutions corrode iron and steel, break down paint films and erode the surfaces of brickwork, stonework and tiles. Acid rain also attacks decorative metal and plastic finishes. Regular cleaning to remove the contamination can prolong the useful life of materials in these environments.

Ageing. Exposure to sunlight can cause bleaching, colour fading of materials and even decomposition owing to solar radiation especially ultra-violet light. Particularly affected are bituminous products, plastics and painted surfaces.

Biological

Timber, including structural, non-structural and timber-based manufactured items are the targets for biological attack. The agents of this are fungi and infestation of wood-boring insects. Given the right conditions an attack by one or both agents is almost inevitable. There are two main types of fungi that cause decay in building timbers, these being dry rot and wet rot.

Dry rot is the more serious and is more difficult to eradicate than wet rot. It is caused by a fungus that feeds on the cellulose found mainly in sapwood (outer layers of a growing tree). This causes timber to lose strength and weight, develop cracks in brick-shape patterns and finally to become so dry and powdery that it can easily be crumbled in the hand. Two initial factors for an attack are damp timber in excess of about 20% moisture content (MC) and bad or non-existent ventilation.

Wet rot does not spread to the same extent as dry rot. It feeds on wet timber (30% to 50% MC) and is most often found in cellars, neglected external joinery, ends of rafters, under leaking sinks or baths and under impervious (waterproof) floor coverings. During an attack, the timber becomes soft, darkens to a blackish colour and develops cracks along the grain. Very often timber decays internally with a fairly thin skin of apparently sound timber remaining on the surface. The recognition, eradication and treatment of timber affected by fungi are covered in Book 1.

Infestation of wood-boring insects commonly called **woodworm**, after the larvae which are able to feed on, and digest, the substance of wood. The female adult beetle lays eggs during the summer months, usually in the end grain, open joints, or cracks in the timber. This affords the eggs a certain amount of protection until the larvae hatch. The larvae then start their damaging journey by boring into the timber, consuming it and then excreting it as fine dust. The duration of this stage varies between six months and ten years depending on the species. During the early spring, at the close of this stage, the larvae bore out a pupal chamber near the timber surface, where they undergo the transformation into adult beetles. This takes a short period after which the beetles bite a way out of the timber leaving characteristic flight holes. The presence of flight holes is often the first external sign of an attack. After emerging from the timber the beetle's instinct is to mate, lay eggs and then die, thus completing one life cycle and starting another. The recognition, eradication and treatment of timber affected by wood-boring insects are covered in Book 1.

Inspections and repairs

Whenever a building firm undertakes major repairs or maintenance it is desirable to undertake a survey of the building. The extent of the measurements, sketches and details taken will depend on the nature and extent of the work. Clearly a survey prior to replacing windows in a house will be very different to that of one involving structural movement. The procedure for carrying out these inspections and surveys is covered in Book 1.

When considering timber repairs a great deal of judgement and negotiation with the client is often required. Can it be repaired cost-effectively or is it cheaper in the long run to replace it? Each job is considered on its own merits, with cost against future service life being the main consideration.

Always wear a dust mask and eye protection when working on, rubbing down or scraping off old paintwork. Surfaces painted prior to the 1960s may contain harmful lead within the paint. In these circumstances, rub down using a wet process (with wet and dry paper) to minimise the potential risks.

Doors

There are many defects associated with doors. Remedial action will depend on the type and location and may range from a simple adjustment through to complete replacement. Table 12.1 and Figure 12.6 cover the most common door defects, causes and recommend remedial action. In all but minor cases consider/discuss with clients the possibility of renewing the door.

Door frames

Defects to external door frames can normally be attributed to either wet rot to the lower end of the jambs or breakout damage in the lock striking plate area as a result of an attempt to force the door. Both of

▲ **Figure 12.6** Remedial treatment to doors

Table 12.1 Door defects

Defect	Sketch	Causes	Remedial action
Will not close, door sticking at head, sill or stile	Stile hits against jamb	• Build-up of paint over time or swelling due to intake of moisture	• Ease (plane) edge or top of door, remove extra material to allow for clearance; repaint/seal exposed edges on completion
Large gaps between door and frame	Large gap	• Shrinkage due to reduction in moisture content	• Add lipping to one or preferably both edges of the door as these should be at least 10 mm thick. It may be necessary to further reduce the door width before fitting them; repaint/seal exposed edges on completion
Sagging, tapered gap at top of door; may be touching floor.	Large gap at top / Touching floor	• Uneven settlement (downwards movement) of hanging stile, wall and door frame • Loose joints in framed doors	• Remove top bevel, lift up door to fit frame; add new piece to the bottom of the door • Dismantle door, re-glue joints and reassemble square
Hinge bound or binding door springs or resists as it is being opened and closed.	Paint buildup / Protuding screwheads / Gap / Distorted stile / Hinge recess too deep	• Hinge recess cut too deep. • Build-up of paint over hinges • Protruding screw heads • Distorted hanging stile	• Pack out hinge recess with a piece of thin card (can be cut from screw box) • Scrape off paint build-up • Remove screws; plug holes and replace with smaller headed screws • Small distortions can be corrected by adding an extra hinge in the centre, otherwise renew door
Twisted door, does not close evenly to stop.	Not closing at bottom	• Door distorted often due to irregular grained timber and variable moisture conditions particularly if not completely painted/sealed • Door frame fixed out of line	• For small gaps, stops can be adjusted or rebates eased to mask situation, otherwise renew door • As for small gaps or re-fix frame in-line

Table 12.1 Door defects *(continued)*

Defect	Sketch	Causes	Remedial action
Decayed or damaged stiles, rails or panel.	Damage or decay	• Absorption of moisture particularly into unsealed end grain and opened joints • Accidental or criminal damage	• Cut away decayed or damaged section and repair using keyed splices and false tenons
Door does not stay latched when closed.	Paint buildup / Latch does not engage	• Build-up of paint over time to the stops or rebate • Distorted door • Defective lock/latch	• Ease stops/scrape off paint build up; adjust striking plate • Rectify distortion and adjust striking plate • Replace lock/latch
Damaged panel	Split in panel	• Shrinkage; accidental or criminal damage	• Cut out panel and stuck (moulded on solid) beads to form new rebate in place of panel groove; fit new panel and secure with planted beads.

these can normally be resolved by splicing in new timber to the area of damage. See Figure 12.7. In the worst cases a new frame may be required. However, this option will cause the most disturbances to the internal plasterwork and decoration.

Windows

Defects to wooden **window frames** are similar to those of doors and doorframes. They will either be associated with poorly fitting opening parts (sashes and casements) or decayed/damaged frames.

These can be rectified using the methods previously outlined for doors i.e. scraping off the paint build-up, easing leading edges, dismantling and re-assembly of sagging casements and splicing of new timber to decayed or damaged areas. In the worst cases the installation of a new window should be considered.

▲ **Figure 12.7** Repairs to doorframes and windowframes

Boxed frame sash windows

This type of window is the traditional pattern of sliding sashes and for many years has been superseded by casements and solid frame sash windows. This was mainly due to the high manufacturing and assembly costs of the large number of component parts. An understanding of their construction and operation is essential, as they will be met with frequently in renovation and maintenance work.

The double-hung boxed window consists of two sliding sashes suspended on cords that run over pulleys and are attached to counter-balanced weights inside the boxed frame.

Figure 12.8 consists of an elevation, horizontal and vertical section of a boxed frame sliding sash window. It shows the make-up of this type of window and names the component parts.

▲ **Figure 12.8** Boxed frame sliding sash window details

Re-cording sashes

The maintenance carpenter is often called upon to renew a broken sash cord (Figure 12.9). It is good practice to renew all four cords at the same time since the remaining old cords will be liable to break in the near future.

The sequence of operations for renewing sash cords is as follows:

1. Carefully remove staff beads.
2. Carefully remove pockets (access pieces cut towards the bottom of pulley stiles).
3. Take out bottom sash. The sash cords should be wedged at the pulley and removed from the groove in the sash.
4. Carefully remove parting beads. Break paint joints first and carefully prise out with a chisel.
5. Take out top sash in a similar manner to the bottom sash.
6. Remove the weights and cords through the pockets. The wagtail will move to one side to give access to the outside weights.
   Note: the weights may not all be the same, so ensure they are returned to their original positions.
7. Thread new cords over pulleys and down to the pockets. A 'mouse' can be used to thread the cords easily. A 'mouse' is a small lead weight that is attached to a 2-m length of string that in turn is tied to the cord. The mouse is inserted over the pulley and drops to the bottom of the frame. The sash cord can now be pulled through. Many carpenters use a length of small chain instead of a mouse.
8. Fasten sash cords to weights. To obtain the length of cord for the top sash, rest the sash on the sill and mark on the pulley stile the end of the sash cord groove. Pull the weight up to almost the top and cut the cord to the position marked on the pulley stile. Wedge the cord in the pulley to prevent the weight from dropping. To obtain the length of cord for the bottom sash, place the sash up against the head of the frame and mark on the pulley stile the end of the sash cord groove. With the weight just clearing the bottom of the frame cut the cord to the position marked on the pulley stile. Wedge the cord in the pulley.
9. Fix sash cords to the top sash and insert the sash into the frame. The cords are normally attached to the sashes by nailing them into the cord grooves. Alternatively the cord can pass through a closed groove and end in a knot.
10. Replace the parting beads. Where these have been damaged new ones should be used.
11. Fix the sash cords to the bottom sash and insert the sash into the frame.
12. Replace the staff beads and check the window for ease of operation. Candle wax can be applied to the pulley stiles and beads to assist smooth operation.

Door and window hardware

The moving parts of locks, latches, bolts and hinges require regular lubrication and cleaning and also need to be kept free from paint build-up in order to ensure their trouble-free operation.

Working from inside,
wedge cords and
remove sashes

Remove cords and
weights through pockets

Cord lengths
for top sash

Cord lengths
for bottom sash

Types of mouse

Cord nailed or knotted

▲ **Figure 12.9** Re-cording sashes

Door furniture

Door furniture can be ruined by incorrect cleaning. It is important to ensure proper care is taken to keep them clean. Dust, contamination and moisture are the main hazards that can affect door furniture. Irreparable damage can be caused to the surface by using proprietary metal polishes, harsh abrasive cleaners or emery cloths on electrolytically deposited or special surface finishes. The correct cleaning procedures for the most common finishes are:

■ *Anodised aluminium finishes* should be dusted regularly, washed periodically with weak detergent solutions and occasionally wiped with wax polish.

■ *Stainless steel finishes* should be dusted regularly and occasionally washed with soap and water.

■ *Plastic products* require only wiping with a damp cloth. Strong sunlight can cause surface deterioration that cannot be easily rectified.

■ *Powder coated finishes* should be cleaned with a soft cloth and any household furniture polish. Industrial solvents must not be used as these can destroy the finish.

■ *Lacquered finishes* (polished brass, etc.) should be cleaned occasionally with good quality wax-free polish. Abrasives and metal polishes should not be used as these will remove the lacquer.

Hinges

A door that is not properly hinged can result in sagging, which can also have an effect on the lock side of the door and frame. An adequate number of hinges are, therefore, always recommended. External doors, heavy internal doors and doors in areas of high humidity should have three hinges. Heavy good quality hinges should always be used on doors where high frequency service is expected.

Tips on the fixing and maintenance of hinges:

■ The leaf with the largest number of knuckles should be fixed to the door frame.

■ The hinge pins must be in true alignment.

■ The hinge recesses must be equal in depth, square and plumb.

■ Check that all screws have been fitted, ensure they are tight and of the correct size.

■ Ensure that metal hinges are lubricated with oil three times a year.

Locks and latches

The correct installation of locks and latches is paramount to their smooth operation. Alignment of the strike plate is essential to ensure correct location of both the latch bolt and deadbolt. Lock and latch cases that are mortised into the door too tightly will result in jamming of the key and bolt.

Tips on the fixing and maintenance of locks and latches:

■ Check that all screws have been fitted and ensure they are tight.

■ Check the alignment of lever furniture and keyhole plates.

- Check the springing on lever furniture and doorknobs.
- Apply graphite to the cylinder if the keys have a tendency to stick in the lock.
- Check the mortising of the lock case and the alignment of the striking plate, where the latch or deadbolt either sticks or will not engage.

Other items such as door closers, floor springs and panic bars, etc., should be checked and maintained in accordance with the specific manufacturer's instructions. Periodically check:

- that all screw fixings are sufficiently tight;
- the operation of the item is correct and smooth;
- adjust and lubricate in accordance with the instructions.

Replacement of hardware

Inevitably items will eventually begin to wear and require replacement. In general these should be replaced with like for like or the nearest alternative. However, proposed replacement does provide the opportunity to upgrade hardware, particularly locks for increased security.

When replacing or upgrading hardware it is often necessary to make good holes and recesses. Figure 12.10 illustrates a typical situation. In general all follow the same procedure:

Old lock removed

May be bevelled to provide key

Filling pieces cut, holes and recesses enlarged

Oversize filling pieces glued and pinned in place

Filling pieces planed off flush, pin holes filled and sanded off flush

◄ **Figure 12.10** Making good holes and recesses

427

1. Cut oversize filling pieces of a similar material. These are often bevelled to provide a key.

2. Mark and cut out enlarged hole and recesses to receive filling pieces.

3. Glue and pin filling pieces in place.

4. Punch in pins, plane off flush, fill pinholes with wood filler, sand off smooth and repaint or seal the area.

Replacement of Door and Window Frames

When extensive repairs are required it is often more cost-effective in the long run to consider a complete replacement. This should be discussed with the building owner. It should be emphasised that although the replacement may be more expensive than the repair initially, the replacement would last a lot longer and also provide the opportunity of upgrading fittings, etc., and modification to suit new requirements.

Before removing frames, always check the wall above for signs of support, see Figure 12.11. Newer buildings may have a concrete or steel lintel, older properties a stone lintel, brick arch, or soldier course or sometimes none at all apart from the frame. Also look out for cracks in the brickwork mortar joints above the opening, which could be indications of structural movement, possibly leading to collapse on removing the frame.

Do not proceed if a means of support is not evident or there are signs of movement. Seek the advice of a structural expert, as arrangements may have to be made for temporary support, the insertion of a lintel and structural repairs to be carried out by others before the frame replacement is made.

Once you are sure the opening is correctly supported, the frame can be cut out and removed in sections in sequence as shown in Figure 12.12.

Concrete seen on face

Metal seen at head

Reinforced concrete 'boot' lintel

Boxed metal lintel

Brick arch or soldier course seen on face

▲ **Figure 12.11** Means of support over openings

▲ **Figure 12.12** Removing, positioning and fixing a frame

When the frame has been removed, clean off any projecting mortar and previous fixings. Make good any damaged brickwork and holes previously occupied by 'built-in' horn fixings. Finally fix in position the new frame.

- Cut off horns and seal cut ends (paint or preservative).
- Place frame in the opening, temporarily holding with wedges at head or sill if required.
- Check head and sill with a level and adjust wedges as required.
- Check jambs for line with a straightedge, plumb one jamb with level, 'sight in' the other, adjust position and wedge as required.

Repairing floors and roofs

If the defect is the result of fungal decay or wood boring insect damage, the procedures covered in Book 1 should be adopted. In these circumstances the details illustrated in the forgoing chapters can be used to replace the components on a like-for-like basis.

Defective joists and rafters in ground, upper floors and roofs can be repaired on a localised basis by cutting away the affected timber and bolting a new piece of timber alongside. Any new member inserted in this way should normally extend to another load-bearing member, such as a wall plate, binder, or purlin, etc., as illustrated in Figure 12.13.

However, where the work is extensive, especially in circumstances where structural timber is affected, like upper floor joists, rafters and purlins, etc., it is often wiser and more cost-effective to have this work referred to a specialist contractor who will be fully equipped and experienced to undertake it.

Floorboards

Where the defect is the result of fungal attack or wood boring insects, the procedure explained before should be adopted. If due to movement (loose fixings), wear or damage, one or more boards can simply be refixed or replaced.

Care must be taken when refixing or replacing floorboards as there is always the possibility of services (water, gas and electricity)

▶ **Figure 12.13** Replacing defective structural members

running below. The surface of the area to be worked on can be scanned before starting work, with a metal/live electric circuit detector. See Figure 12.14. As an added precaution it is wise to turn off all service supplies at the meter/stop valve before any refixing or cutting out operations commence.

Loose boards are best refixed by screwing down into the joists rather than nailing, especially at upper floor levels in older properties where significant amounts of nailing may damage the lath and plaster ceiling through vibration.

1. A jigsaw can be used to cut the ends of a floorboard next to a joist. The ends can then be chopped back to form a bearing (for heading joints) using a wood chisel, or fix a batten to the edge of the joist.

2. Alternatively a small circular saw may be used to cut the heading joints over the joists. The blade should be set to the floorboard thickness.

3. A sharp knife, padsaw or jigsaw can be used along the length of the board to separate the tongue.

4. Punch the fixing nails through the board. Insert two wide blade flooring chisels in one of the edge joints and lever up the board.

5. Where more than one adjacent boards are to be replaced the heading joints should be staggered over different joists.

6. Finally cut the new boards to length and refix in place using a folding technique.

Scan area with detector for services before cutting

Plunge cut with jigsaw next to joist: chop back to form bearing

Chop back or fix batten

Blade guard partially retracted

Leading edge of sole plate up against stop

Temporary stop tacked in position

Alternatively use circular saw to cut heading joint

Flooring chisels used to lever up boards

Replacement board screw fixed

Tongue removed

▲ **Figure 12.14** Removing and replacing floorboards

Stairs

The most common defects encountered are illustrated in Figure 12.15.

Creaking treads

This results from movement between the tread and riser joint when walked on.

■ Where access to the underside of the flight is possible (in cases where it has not been plastered over) the creaking can be remedied by renewing the glue blocks at the tread-to-riser junction and regluing/rewedging the treads and risers in their string housings.

■ Where access to the underside is not possible, gluing and screwing the tread down into the top of the riser can remedy the problem. Small gaps, normally the result of shrinkage, may be apparent at the tread-to-string housing. These should be filled by gluing in thin strips of timber veneer.

Gaps between the wall and string

This may be the result of either shrinkage or movement, or a combination of both.

■ Shrinkage gaps can be masked by the application of a cover mould.

■ Movement gaps are normally the result of the fixings between the wall and string becoming loose or failing, causing the string to move away from the wall when using the stairs. Refix the string back to the wall by plugging and screwing. A cover mould can again mask any gap or damage between the string and plasterwork.

▲ **Figure 12.15** Common stair defects

Damaged nosings

Cutting out the damaged section and splicing in a new piece can repair these using the following procedure:

1. A square length should be let in. Cut at 45° at either end for additional support and a longer glue line.

2. Fix the let-in piece using glue and screws.

3. Finally plane and rub down the let-in piece to match the nosing profile.

4. Where a Scotia mould is used at the underside of the tread-to-riser junction, it is best to renew the whole length, gluing and pinning the new one in place.

Glazing and painting of doors and windows

Maintenance of glazing work involving sealed double-glazing units and large window panes is best undertaken by specialist glaziers, who will be kitted up to undertake the work efficiently and safely.

Reglazing

The carpenter and joiner can undertake reglazing of small single-glazed panes. Wherever possible glass should be pre-cut to size by the glass supplier. Measure the timber rebate sizes and order glass 3 mm undersize in both directions. For example, a piece of glass for a rebate opening size of 150 × 250 mm should be ordered as a cut size of 147 × 247 mm. If necessary, pieces of glass can be cut to size by scoring along the required line using a glass-cutting wheel. Lay the glass on a flat surface with matchsticks placed under the scored line at either end. Apply pressure on both sides to snap in two along the line. Never attempt to cut narrow strips and always wear eye protection and gauntlets.

The first stage is to remove the broken pane. Where practical, sash or casements should be removed from their frame so that broken glass can be removed and replaced with a new piece with relative safety at ground level.

When replacing broken glass at high level ensure that the area below is cordoned off so that no one can enter the area below. Before starting to hack out the broken pane and hardened putty, ensure you are wearing eye protection goggles and gauntlets to protect hands, wrists and lower arms. These should be worn during the whole process as inevitably shards of glass and fine splinters will be created as the pane is removed and the rebates cleaned up.

Start at the top of the pane by removing the old putty or glazing beads with a wood chisel or hacking knife. Remove glazing sprigs (flat or square nails) with pliers and then lever out remaining glass from behind again using a wood chisel. Finally continue hacking out remaining back putty to rebates.

The procedure for reglazing is illustrated in Figure 12.16.

1. Check glass is correct size.

2. Ensure rebates are primed with paint.

3. Work a bead of putty around the back of the rebate.

4. Position plastic seating blocks in the bottom of the rebate to support the glass.

5. Position the bottom edge of the glass on seating blocks. Gently push the glass into the rebate, applying pressure evenly around the edge until a back putty thickness of 1 to 2 mm is achieved.

6. Use glazing sprigs to secure the glass in place. These may be driven in using a pin hammer or the edge of a firmer chisel. (Panel pins should not be used to secure the glass in place, as their round point of contact results in pressure points, which can lead to the formation of cracks.)

7. Work a bead of putty all around the rebate in front of the glass. Only a small bead is required when using glazing beads to secure the glass.

8. Replace glazing beads squeezing putty into glass, or point up the front putty bead using a putty knife or chisel to form a bevelled fillet. Its upper edge in contact with the glass should be just below the rebate sight line. Slight imperfections in the putty fillet can be improved by running over with a wet paintbrush.

9. Trim off the surplus putty to beads and then trim off the back bedding putty on the inside of the glass.

10. Clean the glass to remove oil and putty marks before it dries.

▲ **Figure 12.16** Reglazing procedure

Painting woodwork

Woodwork is painted to provide a decorative finish. However, more importantly it serves also to protect it from the elements. As a carpenter and joiner carrying out maintenance work you may be required to paint newly replaced items as well as repaint existing items that have been eased or repaired. The paint system for wood normally consists of a primer to seal the surface and provide a bond for later coats, an undercoat that provides a smooth opaque covering coat and finally a decorative gloss or satin topcoat.

Preparation

The key to a successful paint system is careful preparation. Paint will not last long on a defective surface.

Bare wood. Normally only needs an initial rub down with glass paper to remove any roughness and sharp arrises. Knots in bare softwood can be full of resin and may later 'bleed' through to the finished paint surface if not sealed. First wipe over the surface with a cloth soaked in white spirit to remove any stickiness and excess resin. Then coat all knots with a knotting solution. On resinous hardwoods that are to be painted, wipe off excess resin using white spirit and seal the entire surface with an aluminium-based wood primer.

1. Apply wood primer to all surfaces and edges, taking particular care to achieve full penetration of any end grain. This is best undertaken prior to fixing in order to ensure full protection. For example, the backs of skirtings, architraves, doorframes and linings etc. are inaccessible when fixed. This is particularly important for external timber.

2. Fill any defects and open-end grain with a wood filler. Always select a waterproof type for external use. Rub down flush with the surface using glass paper.

3. As priming tends to raise the grain of woodwork resulting in a felt-like hairy surface, the whole job will require rubbing down (de-nibbing) prior to over-painting.

4. Wipe off surface with a 'low tack' cloth to remove any surface dust; apply undercoat.

5. When undercoat is dry, apply the topcoat. Refer to paint manufacturer's information with regard to minimum and maximum over-coating times. If the topcoat is applied too soon, the undercoat will tend to bleed into the topcoat causing defects; too long and it may not bond successfully to the undercoat, resulting in early breakdown and peeling off.

Previously painted wood. If in good condition, lightly rub down with glass paper or clean off using a sugar soap solution. This cleans the surface dirt or grease deposits and removes some of the gloss. New coats of paint will not key well on a gloss surface and will easily chip and peel if not rubbed down or cut back.

1. Knot and prime any eased edges or repairs. When dry, rub down to blend in primer to existing paint surface.

2. Fill and rub down any minor defects and imperfections.

3. Remove surface dust; apply undercoat followed by the topcoat within recommended over-coating time.

When the old system has broken down, it is best to completely strip off the old paint, make good and start again from scratch using the same procedure as for bare wood.

Small areas showing signs of deterioration may be repaired without fully stripping the area.

1. Treat any minor areas of soft timber caused by wet rot with a wood hardener.

2. Rub down the surface to remove all loose defective paint.
   Note: Always wear a dust mask when rubbing down paintwork. Surfaces painted before the 1960s may contain harmful lead within the paint. It is best in these circumstances to rub down using a wet process (wet and dry paper) to minimise the potential risk.

3. Apply the paint system as before.

Painting procedure

Internal painting. Protect carpets and furniture with dustsheets. Doors to other rooms can be sealed with masking tape prior to any rubbing down. Open the window to ensure adequate ventilation. This is both for your own health and to help the paint to dry. Always wear a dust mask when rubbing down and eye protection goggles when scraping off.

- Primers and undercoats are applied by brushing out along the grain.
- Topcoats are initially applied along the grain, brushed out across the grain and finally 'laid off' by finishing with gentle brush strokes along the grain. The aim is to produce a thin, even paint film, which does not 'sag' on vertical surfaces or 'pond' on horizontal ones.

External painting. In general the same procedure for internal painting can be adopted, avoiding adverse weather conditions and taking extra care to ensure full paint coverage to avoid the possibility of moisture penetration.

- Do not work in strong sunlight, as this prevents paints drying in the correct way and is likely to cause it to 'blister'. Wait until the area is in shade before painting.
- Do not paint if rain is expected.
- Do not paint first thing in the morning or last thing in the evening, when there might be dew. The resulting moisture will spoil the paint finish.
- Do not paint when there is a risk of frost.
- Do not use paint intended for interior use only.

It is best to aim to paint from mid morning to just after lunch. This will allow the air to dry before starting and the paint film to dry before the early evening dampness starts to form.

Sequence of operations. A logical approach is required when painting framed joinery. The aim is to keep a 'wet edge' blending in adjacent areas of paint, so that joints are not seen when the paint dries.

Figure 12.17 illustrates typical numbered sequences for painting door and windows.

Doors: remove handles, paint in sequence shown. Leading edge should be painted to match the woodwork of the room it opens into

Casement window: paint opening parts before frame and interior sill

Sash window: from inside open sashes as far as they will go, paint all accessible surfaces, reverse sashes and complete painting

▲ **Figure 12.17** Sequence for painting doors and windows

Multi-skill maintenance

Building firms who specialise in maintenance work will often employ or give preference to operatives who possess multiple skills. They will be expected to carry out, in conjunction with their main craft skill, a range of basic skills in other crafts. In addition to painting and glazing of replaced doors and windows covered previously, they may also be expected to undertake minor brickwork, plaster repairs, tiling and other painting tasks.

Replacing bricks

The first task is to attempt to match the pattern, size and make of the original. Measure the brick size and take a small piece to the brick supplier for them to identify. Brick sizes vary slightly due to the way they are made. New metric size bricks are a little smaller than the old imperial ones. When working on older property matching bricks may be difficult to obtain. Try suppliers who specialise in reclaimed building materials. If imperial bricks cannot be obtained, matching metric ones can be bonded into the existing work by slightly increasing the mortar joints.

Figure 12.18 illustrates the procedure to follow when replacing a single brick. In this case the brick previously cut around is the built-in horn of a sill or threshold.

1. Using a bolster and club hammer (do not forget to wear eye protection goggles) cut out the old brick and clean away the old mortar joints. Brush out all dust particles, taking care to ensure nothing is allowed to enter the cavity.

2. Cut the brick to size if required. First gently score a line around the brick using a bolster and club hammer and finally use a heavier blow on the lines to sever it.

3. Prepare a mortar mix, typically 1:6 (1 part Portland Cement to 6 parts bricklaying sand). Sufficient water is added so that the mix has the consistency of soft butter (firm enough not to collapse when heaped, but easily compressed with a shovel). One part lime or a mortar plasticiser may be included in the mix for improved workability. Alternatively, pre-packed bricklaying mortar mixes are available. **Note:** The colour of the sand used should match the existing mortar joints. Red and yellow sands are commonly available.

4. Lay the bed of mortar into the prepared hole.

5. Butter up (apply mortar) to the end and the top edge of the brick. Place it in position and then remove the surplus mortar.

6. Fill any gaps in the mortar joint. When it has started to 'go off', rake the joints out below the brick surface.

Old 'built in' horn position

Cut out to nearest joint

Cut replacement brick to size

Lay bed joint, 'butter' up end and top edge of brick

Place brick in position and clean off excess mortar

Rake out adjacent mortar joints and point up to match existing pointing

Weather pointed

Tooled (bucket handle)

Recess

▲ **Figure 12.18** Replacing a brick

7. After about 24 hours rake out the adjacent mortar joints.

8. Repoint the joints with a mortar mix that matches the original in colour and finish.

Repointing

The most common methods are:

- Weather pointed, which is done with a pointing trowel.

- Tooled, a concave 'bucket handle' finish created by working along the drying mortar with a special jointing tool, or alternatively a metal bucket handle (hence the name) or a piece of 15-mm copper pipe may be used.

- Recessed, created by brushing out the drying mortar with a stiff bristle hand brush; alternatively a more consistent depth can be achieved by working the joint with a piece of timber having a protruding countersunk screw.

Re-laying brick courses

Figure 12.19 illustrates the procedure to follow when relaying or building whole courses of bricks. Refers in this case to under the sill of a reduced-height replacement window.

1. Cut out bricks at either end to form the bond between the adjacent vertical joints.

2. Clean away any old mortar and dust particles.

3. Dry lay the bricks to determine the pattern. Cut the bricks to size if required.

4. Prepare a mortar mix as before, using a 10-mm joint. Approximately 1 kg of mix is required for each brick.

5. Apply a bed of mortar to the existing brick course, approximately 10 mm thick.

6. Furrow the surface to a V-shaped groove with the point of a trowel. 'Butter up' and lay bricks, removing surplus mortar as you go.

7. Repeat the process to lay the subsequent brick courses. Periodically check the bricks are being laid horizontal and in line with a spirit level. Use the end of the trowel to tap the bricks into place if required.

8. Complete the job by raking out the mortar joints and finally pointing them as before.

Repairs to plasterwork

In modern buildings the internal brick and blockwork walls will often have a hard plastered finish. This is normally applied in two layers, a 9 to 12 mm backing coat and a 2 to 3 mm finishing coat. Ceilings and stud partition walls are surfaced with sheets of plasterboard. These may be finished by a 2 to 3 mm coat of board finishing plaster applied onto the plasterboard that acts as the backing. Alternatively the plasterboard joints may be taped up and filled to provide a dry-lined finish, ready for decoration. Dry-line ceilings were often decorated using a textured coating and worked to create a repeating pattern or stipple finish.

In older buildings, the wall plaster may be much softer, but still normally two coats: a thick lime-based backing coat followed by a thin

Area to be bricked up

Cut out bricks at either end to form a bond

Avoid straight vertical joints

Lay bed mortar, butter up and lay bricks. Remove surplus mortar

Periodically check for line and level

▲ **Figure 12.19** Re-laying brick courses

finishing coat. Ceilings and stud partition walls were then finished using **lath and plaster**. This is a system using thin timber laths nailed to the undersides of joists and faces of studs. Wet plaster was pressed up against them and allowed to squeeze between the gaps in the laths forming a key to hold this backing plaster in place. This was finished using a thin coat of finishing plaster.

The external walls of both modern and older buildings may be covered in **rendering**. This is a surface coat of sand and cement mortar applied to a wall for decorative and/or waterproofing purposes.

All of these finishes may crack due to structural movement, damage by accidental impact or be disturbed during renovation work, therefore requiring repairs that the maintenance carpenter and joiner may be asked to undertake.

Patching plasterwork

The first thing to do when patching plasterwork on any background is to protect the floor by covering with a dustsheet. Figure 12.20 shows the procedure to follow when patching a 'blown' or damaged area of plasterwork to a brick or blockwork wall.

1. Tap plaster around the damaged area to 'see' (hear) if any part sounds hollow or loose.

2. Use a club hammer and bolster to hack off all existing loose plaster until the surface is sound.

'Hack off' existing
loose plaster

'Brush off' to remove
dust and loose particles

'Damp down'
with water

Mix plaster
in a bucket

Scoop up plaster from
hawk and apply to wall
in an upward sweep

Reinforce large
areas with repair
mesh or scrim

Comb or scratch area
to provide a 'key'

Brush off and apply
finishing plaster

'Rule off' using a sideways
sawing action,
working upwards

Trowel up to a
smooth finish

Repeat trowelling up whilst
splashing with water

▲ **Figure 12.20** Patching plasterwork

3. Brush down the surface to remove all loose particles and dust, using a stiff bristle or wire brush.

4. Damp down the wall surface by brushing or spraying with water. This prevents the wall 'suction' from drying out the plaster too quickly which could result in cracking on drying. Some surfaces such as concrete, shiny or glazed bricks and impervious engineering bricks do not help the plaster to stick. In these cases brush on a PVC bonding agent before plastering to ensure good adhesion.

5. Add a small amount of backing plaster into clean cold water in a bucket. Stir with a timber stick until a thick creamy consistency is achieved.

6. Transfer some of the plaster to the hawk. With the hawk tilted away from the wall scoop up a small amount of plaster on the edge of the steel trowel and press the plaster against the wall using an upward sweep of the trowel. The trowel should be used at an angle to the wall, with the angle reduced as you sweep it up the wall.

7. Take care not to allow the trowel to lay flat against the wall, as the suction will pull the fresh plaster off the wall.

8. Continue adding plaster to the wall until the whole area to be repaired is covered to within 2 to 3 mm below the surrounding wall finish. Pushing into the backing coat a repair mesh or scrim will reinforce larger areas.

9. Before the backing plaster is completely set, scratch the surface with a comb. This provides a key to help with the adhesion of the finishing coat.

10. After about 3 to 4 hours the backing coat surface will be hard, but not dry. It is then ready for finishing. If allowed to dry further, it will require damping down again with water before finishing.

11. Brush down the surface to remove any loose particles. Mix up a small batch of finishing plaster by adding to a little water as before, except this time it should be a runnier consistency. Trowel on the plaster, aiming to leave it slightly proud of the surrounding area.

12. Rule flat the surface using a timber or metal straightedge. Start at the bottom of the patch, move it up the wall with a side-to-side sawing action keeping it tight against the existing sound plaster as a guide. Trowel on more plaster to fill any hollows before ruling off again.

13. The plaster will start to set within 30 to 45 minutes. At this stage smooth the surface with a plastering trowel, again using upward sweeps with the trowel held at an angle. After 15 or so minutes lightly splash the surface with clean cold water, whilst trowelling up and over the surface, to provide a smooth hard finish. Ensure the trowel is kept clean and damp during this process to prevent damaging the newly plastered surface. Regular brushing off in a bucket of cold water is ideal.

14. Small patches in rendering can be repaired in one or two coats using the above procedure, except that a sand and cement mix is used in place of plaster.

Patching damage to plasterboard or lath and plaster surfaces

To repair small holes caused, for example, by striking the surface with the corner of a piece of furniture, the procedure to follow is illustrated in Figure 12.21.

1. Neaten up the jagged edges using a sharp knife for plasterboard or a pad saw for lath and plaster.

2. Cut a strip of plasterboard about $1\frac{1}{2}$ times the length of the neatened hole and just narrower in width. Make a hole in its centre, pass through a piece of string and knot it behind on a nail. (Remember plasterboard is simply cut by scoring on the face with a sharp knife, break along the line by applying pressure along from the scored side. Run the knife along the paper on the other side to separate.)

3. Mix up some plasterboard adhesive and apply to both ends of the strip. Feed the strip into the hole, using the string to pull it tight against the inner face of the board or laths. Tie off the string to a scrap of timber positioned over the face of the hole.

4. When the adhesive has set, cut off the string. Cut another piece of plasterboard, this time to fit the hole, and again bond in place using plasterboard adhesive. Press in until it is just below the surrounding wall surface and leave to set.

5. Fill the patched hole using board finishing plaster and trowel up as before.

Repairing large holes

For large holes caused, for example, by a foot slipping through the ceiling when working in a loft, the procedure to follow is illustrated in Figure 12.22.

1. Mark on the ceiling two lines at right angles to the joist direction and enclosing the damaged area.

Neaten up hole

Strip of plaster board to bond behind hole

Cut strip with sharp knife

Apply adhesive to strip-ends and hold in place

Bond piece of plaster board to strip. Fill flush with board finishing plaster

▲ **Figure 12.21** Patching small holes in parition walls

Square up damage
between joists

Battens fixed to
sides of joists

Use of screw fixings
cause far less
disturbance

Noggin fixed
between joists

cut and fix piece
of plasterboard

▲ **Figure 12.22** Repairing large holes in plasterboard

2. Use a pad saw to cut along these lines until the adjacent joists are reached after first checking for the presence of cables and plumbing.

3. Again using the pad saw cut along the joist edges. Remove the damaged area, leaving a neat rectangular hole.

4. Cut battens; fix to the sides of the joists.

5. Cut and fix noggins between the battens at either end of the hole, ensuring the noggins' centre lines straddle the cut line, to provide a bearing for both the existing sound ceiling and new plasterboard.

6. Cut a piece of plasterboard 2 to 3 mm smaller than the hole in both directions. Fix in place into the noggins and battens using plasterboard nails or plasterboard screws.

7. Use a sharp knife to cut away the finishing plaster about 25 mm all round the hole. Bed lengths of plasterer's scrim over the joints between the patch and existing sound ceiling, using a thin (runny) mix of board finishing plaster as an adhesive. This is to reinforce the joint and reduce the risk of subsequent cracking.

8. Fill the patched area using board finishing plaster and trowel up as before.

Damaged reveals

To make good the plaster work around the reveals of a replaced door or window: in most circumstances these can be patched using the two-coat backing and finishing plaster method as before.

Where there is extensive damage, or the plaster is loose, the entire reveal should be hacked off and replaced. Figure 12.23 illustrates the procedure to follow using plasterboard and board finishing plaster:

1. Hack off plaster reveal back to the brick or blockwork surface and extending around the corner by about 75 to 100 mm.

2. Cut two strips of plasterboard, one for the reveal and the other for the return.

3. Brush down the wall surfaces. Mix up some plasterboard adhesive. Using a trowel or special caulker, apply dabs of adhesive up the centre of the reveal and continuously around the perimeter.

4. Press plasterboard strips in place and check for plumb with a spirit level. The return strip should finish 2 to 3 mm below the adjacent plaster to allow for a coat of board finish.

5. Reinforce the corner with a length of metal plasterboard bead, which also acts as a guide for the subsequent finishing plaster coat. Apply a bed of plasterboard adhesive or board finish to the corner. Press the bead in place, check with spirit level for plumb. Also ensure it is in line with existing wall surface. Remove excess adhesive and allow too dry for 2 to 3 hours.

6. Reinforce the joint between return plasterboard strip and existing wall plaster using a length of scrim as before.

7. Complete repair by applying a coat of board finishing plaster and trowel up in the normal way.

▲ **Figure 12.23** Making good plasterwork to damaged reveals

Replacing damaged tiles

Probably the hardest task is to find a good match for replacement. The building owner may have spares left over from the original work. Alternatively, take a damaged piece to a tile supplier for them to find a match. Figure 12.24 illustrates the procedure to follow:

- Prepare the surrounding area and yourself. Cover the floor and surrounding units or bath and sanitary ware with dustsheets to protect from dust and possible scratching by the small sharp particles of broken tiles. Protect yourself by wearing eye protection goggles, gloves and a dust mask.

- Rake out grout joint around the damaged tile to relieve the perimeter stresses.

- Drill a series of holes around the centre of the tile and break out using an old chisel, working progressively towards the edges. Do not try to break out the tile by trying to prise it off from the edge joints, as almost inevitably you will damage the adjacent tiles. Masking tape can be applied when drilling out the centre to prevent the masonry drill skidding across the ceramic surface.

- Scrape or chip off the old tile adhesive back to the surface, taking care not to damage the plaster base.

- Apply four dabs of tile adhesive to the back of the tile, or use a notched comb to provide a uniform ribbed layer.

- Press the tile in place so that it lies flush with the surrounding ones. Insert tile spacers or matchsticks in the joints to position or support the tile. Adjust the tile as required to ensure a uniform gap all around.

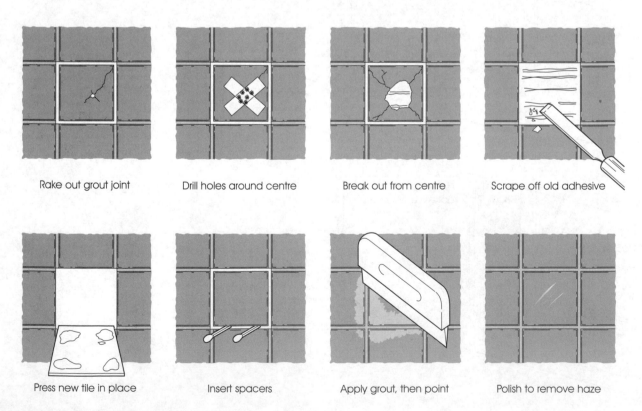

Rake out grout joint   Drill holes around centre   Break out from centre   Scrape off old adhesive

Press new tile in place   Insert spacers   Apply grout, then point   Polish to remove haze

▲ **Figure 12.24** Replacing a damaged tile

- Allow the adhesive to set for about twenty-four hours and then remove the spacers.

- Fill the gap around the tile with grout, working in with a rubber squeegee. Point the joint with a fingertip or piece of wood dowel with a rounded end point.

- Finally when dry, polish up the surface to remove the grout haze using a clean dry cloth.

The long-term success of tiling depends on the adhesive and grout used. Most are available either in ready-mixed or powder form for mixing with water to a creamy paste. A standard type mix is only suitable for dry areas. It may also tolerate a little condensation or occasional splashing with water. In areas subject to more prolonged condensation or extensive wetting such as a shower area, always use waterproof products.

Cutting tiles

Tiles are best cut using a proprietary tile cutter, or diamond-tipped wet saw. However, small amounts can be cut by scoring the surface with a carbon-tipped tile-cutting point and snapping the tile along the scored line. Place two matchsticks under the scored line. Press down firmly on either side to snap in two, see Figure 12.25. Corners and curves can be cut out of tiles to fit around projections using a tile saw blade in a coping saw frame. Alternatively the lines of the area to be removed may be scored and the waste nibbled away with a pair of pincers.

Finishing off

The joint between tiles and horizontal surfaces such as kitchen worktops, baths and sanitary ware will require sealing with a silicone sealant to prevent moisture penetration. Figure 12.26 illustrates the procedure to follow to ensure a neat bead of sealant in these locations.

- Apply masking tape to both the vertical and horizontal surfaces, 2 to 3 mm away from the internal angle. Trim the cartridge nozzle off at an angle of 45° to give a bead just wide enough to fill the gap between the two taped edges.

- Gently squeeze the cartridge gun trigger until the sealant is just seen at the tip.

- Place the nozzle at one end of the angle to be filled, with the gun held at 45° to the wall.

Score along line with
tile cutting point

Place matchsticks under
score, press down to snap
tile in two

◀ **Figure 12.25** Cutting ceramic tiles

Apply masking tape
along angle to be sealed

Push the cartridge nozzle
along the angle

Allow sealant to skin
over, peel off tape

Complete
set bead

▲ **Figure 12.26** Sealing tiles to a worktop

- Apply steady, even pressure to the trigger whilst pushing the gun along the angle. The nozzle will form the sealant into a neat concave curve. To stop the flow at the corners and on completion, release the metal tag adjacent to the trigger.

- Any unevenness in the bead can be smoothed using a small paintbrush dipped in water. Leave sealant for a short while to skin over, and then peel off the tape to leave a well-formed neat bead.

Painting plasterwork

Walls and ceilings are normally painted using emulsion paint. Newly plastered and repaired surfaces should be left for seven to ten days to dry and then treated with a coat of plaster sealer before decoration. This prevents the new plaster showing through the paint finish in the form of 'patchiness'. Alternatively, a thinned emulsion can be applied as a primer before at least two full-strength coats are applied. The priming coat should be about one part water to about three parts emulsion paint.

Before starting work, arrange for any furniture in the room to be removed. Protect carpets and fixtures with dustsheets. Wear a dust mask and eye protection goggles when rubbing down and scraping off.

The procedure to follow is illustrated in Figure 12.27:

1. Remove all loose material such as dirt, dust and flaking paint.
2. Rake out any minor cracks in the plaster surface, using the end of an old slot blade screwdriver. The raking out is to make the crack a little deeper and wider with undercut edges.
3. Fill cracks with plaster filler. Ensure filler is pressed well into the cracks in order for it to key on the undercut edges.
4. Leave filler slightly proud of surrounding surfaces. Rub down level when dry. Fill the area again and rub down if required.
5. Rub down the entire wall or ceiling surface.
6. Wash down the surface and allow it to dry.
7. Ceilings should be painted before walls. A small brush is used to cut into the corners and up to the frames and skirting, etc.

Painting a wall

Painting a ceiling

Rake out cracks, undercut edges

Fill cracks, leave filler proud of surface

Rub down level, repeat fill and rub down if required

▲ **Figure 12.27** Painting plasterwork

8. Use a roller or large brush to cover an area of about 1 m^2 at a time. Apply paint in one direction; spread it out by brushing or rollering diagonally. Finally finishing off using light pressure only, in the same direction as you started.

9. Using the numbered sequence continue painting the subsequent squares or strips, blending in the paint application of one with another whilst the paint is still wet; otherwise if wet paint is applied over a drying one, pronounced lines will be apparent in the finished work.

10. Clean all brushes/rollers and equipment in water on completion.

Index

Abutment 37
Access panels 316
    traps 28
Ageing of materials 418
Anchors, framing 89
    joists 39
    roofs 53
Angle tie 49
Arch construction 337
Architraves, types 262
    installing 265
Attic rafters 66

Balloon framing 105, 109
Balusters 131
Balustrades 131, 135, 149
Bargeboards, roof 74
Barriers, cavity 117
    vapour/moisture 115
Bars, counter 296
Bath casings 316
Battering, foundations 400
Bay windows 231
    roofs 235
Beads, door panel 165
    glazing 173
Beams, box 349
    concrete 383
    cramp 349
    formwork 383
    glulam 122
    joist joints 15
Bench seating 307, 309
Binders, joist 14
Biological attack, materials 419
Blinds, Venetian 246
Block wood flooring 33
Blow hole, formwork 349
Boarding, cladding 118
    flat roofs 42
    flooring 25
    laying 26
Bolection moulding 166
Bolts, hold-down 368
    security 195, 204

Bow windows 234
Boxed frame, units 280
    windows 244, 250
Box-outs, formwork 349, 377
Bracing, roofs 64
Brick cladding 118
Brickwork, maintenance 437
Building regulations, stairs 133
Buildings, knock-down 106
    site huts 106
    timber framed 108
Bull's eye window 239
Bull-nosed steps 131, 135

Cantilever joist 19
Capillarity 412
Carcassing 279
Carriage, stairs 131, 144
Casement windows 222, 225, 236
    assembly 241
    fixing 257
    French 231
    linked 254
Casings 313–319
    access panel 316, 318
    bath 316
    ladder frame 314
    pipe work 315
    services 313
Cavity barriers 117
Ceiling, exposed grid 322
    false 319
    joists 47, 54
    levelling 323
    panels 321
    suspended 319
    Z-bar 323
Cement rendering 118
Centring for arches 337
Chemical attack, materials 418
Chimney stack trimming 52, 66, 68

Chipboard, flooring 25, 29
    formwork 360
Church furniture 308, 311
Cladding 118, 333
    grounds 334
    materials 335
    venting 334
Column cramp 350
Columns, formwork 349, 370
Comb joint 229
Combination frames, door/window 230
Concrete, floors 5
    formwork 350
    stairs 389
Condensation 416
Constructions, temporary 337–409
Cornice 263
Counters, bar 296
Covering, floor 25
Cramp, beam 349
    column 350
    floorboard 26
Cripple rafters 47
Cube test, concrete 350
Cupboards 279, 288, 291
Curtail step 136
Curtain walling 106, 125
Cylinder rim lock 203

Dado rail 262, 277
Damp problems 412
    rising 415
Deadlocks 195
Decking 28
    flat roofs 42
    formwork 350
    span 9
Decorative panels, doors 166
    walls 324
Defects, causes 411
    doors 421
    joists 12
Design failures 411

Dome roofs 86
Doors 154–221
    assembly 162
    combination frame 230
    concealed closure 194
    cupboard 288, 303
    curved 177, 209
    decorative panel 166
    defects 421
    double margin 175
    double-faced 176
    dowelled 167
    fanlight 208
    fire resistance 156
    fire-resisting 179
    flush 157, 171
    folding 182, 200
    frames 155, 207, 214, 420, 428
    furniture fixing heights 198
    garage 160
    glazed 157, 169
    Gothic 161
    grading 158
    handing 188
    hanging 183
    hingeing 183
    hollow 171
    honeycomb core 172
    industrial 160
    insulation 156
    ironmongery 188
    joints 167
    knobset 195
    lattice 172
    lining 155, 210, 220
    lippings 173
    lobbies 214
    lock block 172
    locks/latches 194, 203
    louvred 178
    maintenance 420, 426
    matchboarded 157, 158
    muntins 165
    openings 101, 103
    operation methods 181
    overhead closure 193
    panelled 157, 164, 169
    performance 156
    pivoted 191
    privacy 156
    rails 165
    repair 420
    replacement 428
    revolving 182
    safety glazing 174
    seals 193
    security 156
    sets 212
    sizes 158
    skeleton core 171
    sliding 161, 199
    solid core 172
    stiles 165
    swinging 182
    transom joints 205
    types 157
    up-and-over 182
    vision panels 173
    warehouse 160
    weatherboard 186
Doorstop 195
Dormer windows 69, 82
Double glazing 252
Dowelled joints 167
Draft seal, window 229
Dragon tie 49
Drawers 282, 303
Drawings, scale 55
Drip batten, roof 43

Eaves 37, 43
    finishing 76
Encasing services 313
Entrance lobbies 214
Excavations, timber supports 399

Facings, stairs 152
Failure causes, materials 411
False ceilings 319
Fanlight, door 208
Fascia boards, roof 75, 77
Faults, doors 421
Fences 402
    post sockets 407
    posts 406, 409
    preservation 408
Fibreboard, formwork 360
Fillet, formwork 352
Finishing, flooring 32
    stairs 144
    verges/eaves 73, 76
Fire door seals 179
Fire protection, joists 14, 20
Fire resistance, doors 156, 179
    partitions 98, 104
Fire stops 117

Fireplace joists 21
    openings 28
Fixings to foundations 112
Flat pack units 301
Flat roofs 38–42
    decking 42
    metal covering 42
Floorboards 25
    laying 26
    tightening 27
Floors 1–34
    beam and block 2
    boards 25
    concrete 5
    construction 6
    covering 25
    decking 28
    double/triple 4
    finishes 32
    framed 4
    ground 1
    insulation 29, 33
    joists 6–24
    laying 26
    preservation 31
    repairs 430
    services 27
    sheet 28
    solid 1
    suspended 2, 4
    T&G 25
    types 1
    upper 4, 110
    vapour barrier 33
    wood blocks 33
Flue trimming 52
Folding doors 182, 200
Folding wedges 97
Form ties, formwork 353
Formwork 349–391
    beams 383
    columns 370
    concrete 350
    construction 365
    curved 380
    excavations 399
    foundation 366
    kicker 355, 366, 368
    loading 363
    metal 361
    panels 353
    plastic 362
    release agent 362
    safety 395

stairs 389
template 372
textured 360
timber 349, 358
walls 376
Foundation formwork 366
Foundations, fixings to 112
Fox wedge joint 168
Frames, anchors 89
ceiling 321
combination door/window 230
door 155, 207, 214, 420, 428
panelling 326
plywood 123
squaring 162
timber 105–120
winding 162
window 222–254, 422, 428
French casements 231

Gable ends 36, 49, 65
erection 60
ladder 65
Gambrel roof 37
Gang mould, formwork 353
Garage doors 160
Gate hinges 408
Glass, low emissivity 252
Glazed doors 157, 169
screens 175, 214
Glazing, beads 173
repairs 433
double 252
secondary 253
windows 151, 222
Glulam timber 106, 121
Gothic arch construction 338
doors 161
Grades of timber 88
Grading, doors 158
Ground movement 417
Grounds, cladding 334
panelling 328
skirting 273
Guard rails, stair 134
Gutter, boxed 44
parapet 51

Handing of doors 188
Handrail, stairs 134, 137, 143, 150
Hangers, joist 13

Hanging doors 183
Hatch trimming, roofs 51, 66, 68
Head plate, partition 90
Heading joints 26
Headroom, stair/landing 134
Heat protection, joists 20
Hinge bolt 195
Hingeing doors 183, 186
Hinges, door 189
gate 408
maintenance 426
recessing 186
window 255
Hip rafter 46
Hipped-end roof 36, 49
erection 61, 66
Hoardings 402
Honeycomb core doors 172
Huts, site 106

Industrial doors 160
Inserts, formwork 355
Inspection, buildings 420
formwork 395
Insulation, doors 156
floor 29, 33
partitions 98, 104
roofs 44, 79
site hut 107
timber walls 116
Ironmongery, doors 188
windows 255

Jack rafters 59
Jerkin-head roof 37
Jig, panel 114
Joints, door 167
fox wedge 168
glulam 122
joists 17
mortise and tenon 168
partitions 88
skirting 268
stress 18
windows 227, 244
Joist hanger 18
Joists
beam joints 15
binders 14
cantilever 19
ceiling 47, 54
cutting 20
defects 12

fire protection 14
floor 6–24
hangers 13
heat protection 20
jointing 17
layout 23
layout, flat roof 38
levelling 21
notching/drilling 24
positioning 19
preservative 31
restraints 15
sections 11
services 24
sizes 6
spacing, floors 8
spacing, roofs 54
span, floors 7
span, flat roofs 40
strutting 15
supports/anchors 39
supports/restraints 12
trimming 17
wall plates 13

Keyhole escutcheon 196
Kicker, formwork 355, 366, 368
Kicking plates, door 199
Knobset, door 195
Knock-down buildings 106
units 280
Knots, timber 12

Ladder frame, roofs 47, 65
casings 314
Landing headroom 134
stairs 144
Latches, door 194, 203
Lath and plaster repairs 441
Lattice doors 172
Lean-to roof 36
Lectern 311
Letter plates 199, 204
Linings, door 155, 210, 220
formwork 361
wall 107
Lippings, door 173
Loading, formwork 363
Lobbies 214
Lock block, doors 172
Locks, door 194, 203
maintenance 426
windows 255

Loft conversion, stairs 135, 140
Louvred doors 178
Low emissivity glass 252

Maintenance 410–449
    brickwork 437
    doors 420, 426
    locks/latches 426
    paint 433
Mansard roofs 36, 82
Matchboard panelling 332
    doors 157, 158, 164
Materials, ageing 418
    biological attack 419
Metal stud partitions 99
Metal-covered roofs 42
Mitres 265, 269
Moisture barriers 115
Mortise and tenon joints 168
Mortise lock 195, 201
    rack bolt 205
Mouldings/trim 262–278
    types 263
    cutting 264
Moulds, panels 165, 332
    pre-cast concrete 392
Mullions 247
    intersecting 233
Muntins, door 165

Nail plates 63
Nailing schedule, frames 113
Newel, stairs 131, 137, 143
Noggins, partitions 92
Nosing, stairs 131
Notching joists 24
    partitions 98

Ornamental roofs 84
OSB flooring 25

Padlocks 194
Paint maintenance 433
Painting plasterwork 448
    sequence 437
Panelling 324–333
    grounds 328
    external 113
    formwork 353
    insulation 116
    jig 114
    matchboard 332
    moulds 332
    partitions 101

Panelled ceiling 321
    doors 157, 164
Panels, access 316
    door 165
    stressed skin 120
    vision, door 173, 181
Panic bolts 194, 197
PAR section timber 88
Parapet gutter 51
    roof 37
Park bench seating 309
Partition walls 87–104
    construction 89
    door openings 101, 103
    fire resistance 98, 104
    head plate 90
    insulation 98, 104
    jointing 88
    metal stud 99
    noggins 92
    notching/drilling 98
    openings 92
    panels 101
    pre-made 97
    services 97, 103
    studding 88, 110
Pew, church 308
Picture rail 263, 277
Pipe work, casing 315
    in partitions 97, 103
    under floors 24, 27
    under skirting 275
Pitched roofs 35, 46
Pivot windows 222, 237, 239
Plasterboard 88, 94
Plasterwork, repairs 439
Plates, nail/wall 63
Platform frames 105, 108
Platform, water tank 53, 66
Plinth block 262, 267
    cupboard 280
Plywood, formwork 359
    frame 123
Posts, fence 406, 409
    sockets 407
Pot board 279
Pre-cast concrete moulds 392
Prefabricated roofs 62–73
Preservation, cladding 336
    fences 408
    roof timbers 79
    joists 31
    wall timbers 115
Pressure, formwork loading 365

Privacy, doors 156
Props, adjustable 349
Protection, weather 156, 334
Purlin, roof 47

Quadrant moulds 268

Rafters 46
    attic 66
    calculations 59
    erection 70
    setting out 57
    sizes 54
    trussed 62–73
Rails, door 165
Rain penetration 412
Reglazing 433
Regularised timber 11
Regulations, stairs 132
Release agent, formwork 362
Rendering 118
Repairs/maintenance 410–449
    brickwork 437
    floors 430
    glazing 433
    lath and plaster 441
    plasterwork 439
    reveals 445
    roof 430
    stairs 432
    tiles 446
Restraints, joist 15
Reveals, repair 445
Revolving doors 182
Ridge, roof 37, 81
Ridges, roof 81
Rim lock 195
Risers, step 131, 136
Riser/tread details 136
Rising damp 415
Roofs 35–86
    anchoring 53
    angles 56
    attic 66
    bargeboards 74
    bay windows 235
    bracing 64
    dome 86
    dormer 69, 82
    double 48
    drip batten 43
    eaves 37, 43, 76
    elements 37, 46
    ends 49, 64

erection 60, 70
fascia boards 75, 77
flat 38–42
gable ladder 65
gutter, boxed 44
insulation 44, 79
irregular plan 81
ladder frame 47
leaking 412
Mansard 82
ornamental 84
pitch 35
pitched 46
prefabricated 62–73
preservation 79
rafter 46
repairs 430
ridges 81
shapes 36
shell 123
single 47
size allowances 58
sizes 53
soffit boards 79
spacing 53
spans 54
ties 44, 49
trimming 51, 66, 68
triple 49
trusses 83
turret 86
types 35
valley 38, 50, 66
ventilation 44, 79
verges 43, 73

Saddle board 49
Safety glazing, doors 174
Safety, formwork 395
  shoring 348
  striking 399
Sash windows 223, 248, 259
  repairs 423
Sashes, balancing 248
  re-cording 424
  sliding 244
Schedule, nailing 113
Scotia moulds 268
Screens, glazed 175, 214
Screens/partitions 298
Seals, door 193
  fire doors 179
  window 229, 252

Seating 305
Secret fixing boards 25, 27
Sections, joists 11
Security bolts 195, 204
  chain 195
  doors 156
Services access 28
  in joists 24
  in partitions 97, 103
  encased 313
  under floors 27
  under skirting 275
Sheet flooring 28
Shell roofs 106, 123
Shelves, adjustable 281
Shiplap boarding 118
Shoring 340–349
  removal 348
  safety 348
  timber sizes 348
Shuttering 357
Sills, window 244
Skirting 262
  grounds 273
  installing 268
Slabs, roof/floor 382–388
Sliding doors 161
  ironmongery 199
Sliding windows 244, 250
Soffit boards 78
  formwork 357
Sound insulation, doors 156
  floors 30
  partitions 98, 104
Spacing, stud 91
Span, decking 9
  flat roof joists 40
  floor joists 7
  roofs 54
Spandrel, stairs 131
Specification, doors 156
Spreader plate 47
Squaring of a frame 162
Stairs 128–153
  balusters 131
  balustrades 149
  carriage 144
  concrete 389
  construction 132
  facings 152
  finishing 144
  geometrical 130, 141, 147
  guard rails 134

handrails 143, 150
headroom 134
landing 144
loft 135, 140
newel 131, 137, 143
open plan 139
open-riser 147
paddle-tread 141
pre-cast 394
protection 152
regulations 132
repairs 432
spine-beam 147
steps 131, 135
string assembly 137
string fixing 143
tapered 145
terminology 130
types 128
wreathed string 142
Steps, bull-nosed 135
  stairs 131, 135
  tapered 145
  tread/riser 136
Stick-built construction 114
Stiles, door 165
Stormproof windows 227, 238
Stress, joints 18
Striking, formwork 398
  safety 399
  strip 384, 398
  times 398
String, stairs 132
Strut, roof 47
Strutting, joist 15
Stud partitions 88
  spacing 91
Subsidence 417
Suspended ceilings 319
  floors 2, 4
Swinging doors 182

Table construction 312
Table form, formwork 358
Template, formwork 372
  stairs 132
Temporary constructions
  337–409
Thermal insulation, doors 156
  floors 29, 33
  partitions 98, 104
  roofs 44
  trussed roofs 79

Thumb latch 197
Ties, roof 47, 49
Tile cladding 118
Tiles, repair 446
Timber, glulam 121
    grades 88
    preservation, walls 115
    regularised 11
    sizes, shoring 348
Timber-frame construction
        105–120
    buildings 108
Tongued and grooved flooring
    25
Transom joints, doors 209
    windows 227
Treads, stairs 132, 136
    tapered 145
Trim/mouldings types
        262–278
    cutting 264
Trimming joists 17
    to chimneys 52, 66, 68
    to openings 51
Trussed rafters 62–73
Turret roofs 86

Units/fitments 279–305
Up-and-over doors 182
Upper floors 4, 21, 31, 110

Valley rafter 47
    roofs 38, 50, 66
Vapour barriers 33, 115
Venetian blinds 246

Ventilation, roof 44
    trussed roofs 79
    window 240
Ventilator, hut 107
Venting, cladding 334
Verges, roof 38, 43
    finishing 73
Vision panels, door 173, 181

Waffle form, formwork 358
Waling, formwork 358
Wall plates 13, 46, 63
    units 302
Walls, curtain 106, 125
    curved 380
    damp 412
    external 110, 113
    formwork 376
    insulation 116
    partition 87–104
    studding 89
    timber preservation 115
Wardrobes 291
Warehouse doors 160
Water tank platform 53, 66
Weather protection, 145, 334
    protection, doors 156, 186
Wedges, folding 97
Winding of a frame 162
    strips 163
Windows 222–261
    assembly 241
    balancing sashes 248
    bay 231
    bow 234

    boxed frames 244, 250
    bull's eye 239
    casements 222, 225, 236
    combination frame 230
    curved 247
    dormer 69, 82
    double glazed 252
    features 223
    fixing 257
    frames 222–254, 422, 428
    French casements 231
    glazing 222, 251
    hinges 255
    hut 107
    ironmongery 255
    joints 227, 252
    linked 254
    locks/fasteners 255
    mullions 231, 247
    pivot 222, 237, 239
    repair 422
    replacement 428
    sash 223, 248
    seals 229
    secondary glazed 253
    sills 244
    sliding 244, 250
    special 238
    specification 224
    stormproof 227, 238
    types 222
    ventilation 240
    window boards 260
Wood blocks, flooring 33
Worktops 290, 303